International Investment and Climate Change:
Energy Technologies for Developing Countries

Research by the Energy and Environmental Programme of the Royal Institute of International Affairs is supported by generous contributions of finance and technical advice from the following organizations:

Amerada Hess
Ashland Oil
British Gas
Blue Circle Industries
British Nuclear Fuels
British Petroleum
Eastern Electricity
ENI
Enron
Enterprise Oil
Esso/Exxon
LASMO
Mobil Services
Mitsubishi Fuels
Osaka Gas
PowerGen
Ruhrgas
Saudi Aramco
Shell
Statoil
Tokyo Electric Power
Texaco
Veba Oil

International Investment and Climate Change:
Energy Technologies for Developing Countries

Tim Forsyth

Energy and Environmental Programme

Earthscan Publications Ltd, London

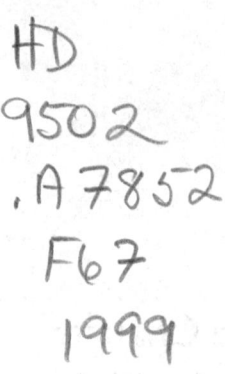

HD
9502
.A7852
F67
1999

First published in the UK in 1999 by
The Royal Institute of International Affairs, 10 St James's Square, London, SW1Y 4LE
(Charity Registration No 208 223)
and
Earthscan Publications Ltd, 120 Pentonville Road, London N1 9JN

Distributed in North America by
The Brookings Institution, 1775 Massachusetts Ave NW,
Washington, DC 20036-2188

Copyright © Royal Institute of International Affairs, 1999

All Rights Reserved

A catalogue record for this book is available from the British Library

ISBN: 1 85383 622 2 paperback

Typesetting by Composition & Design Services, Minsk, Belarus
Printed and bound by Thanet Press Limited, Margate, Kent
Cover Design by Yvonne Booth

The Royal Institute of International Affairs is an independent body which promotes the rigorous study of international questions and does not express opinions of its own. The opinions expressed in this publication are the responsibility of the author.

Earthscan is an editorially independent subsidiary of Kogan Page Limited and publishes in association with WWF-UK and the International Institute for Environment and Development.

This book is printed on elemental chlorine free paper.

Contents

Foreword by Michael Grubb .... *x*
Preface .... *xiii*
Acronyms and abbreviations .... *xiv*
Executive summary .... *xviii*

Part I Themes

1. Introduction **3**
   International investment and climate change 3
      Environmental policy and renewable energy development .. 4
      Technology investment and privatization 6
      Climate change policy and technology transfer 9
   Aims and structure of the book 12

2. International investment and climate change mitigation **16**
   Introduction 16
   The attractions of harnessing foreign investment 17
      The growth of the private sector 17
      Privatization and liberalization of electricity supply industries 19
   Utilizing the firm in public and environmental policy 21
      The firm as an economic unit 22
      Privatization and public–private synergy 23
      Global investment and technological competitiveness 25
      Business investment and environmental policy 28
   International investment and climate change policy 33
      The Joint Implementation controversy 33
      The experience of Activities Implemented Jointly 36
      The Clean Development Mechanism 37
      Enhancing international investment for local development 39
   Summary 41

3. Decentralized electrification and climate technology transfer — 43
Introduction 43
Decentralized electrification and renewable energy 44
    Defining renewable energy technologies 45
    Rural electrification and renewable energy 47
Electricity investment and technology transfer 51
    Defining technology transfer 51
    Improving technology transfer in practice 56
Integrating renewable energy and technology transfer for
climate change mitigation 59
    Technology transfer under the UNFCCC 59
    Institutional and financial assistance for renewable
    energy development 63
Summary 68

4. Electricity investment and privatization in South-east Asia ... 70
Introduction 70
Electricity investment and private investment 72
    Trends in electricity demand and supply 72
    The growth in IPPs 75
Opportunities for renewable energy development 78
    Renewable energy resources in South-east Asia 78
    Opportunities for decentralized electrification 80
The Asian financial crisis 83
    The nature of the crisis 83
    Opportunities and implications for electricity investment 85

Part II Case Studies

5. Introduction to the case studies 91
Introduction 91
Selection of case studies 93
Structure and aims of case studies 96

6. Building renewable energy in grid-dominated areas: the experience of Thailand 100
Introduction 100

Energy supply and renewable energy resources in Thailand... 101
   Overview 101
   Energy trends 102
   Renewable energy resources 104
Structure and liberalization of the electricity supply industry 105
   Electricity market and institutions 105
   Privatization and liberalization 110
Renewable energy and technology transfer 115
   Vertical transfer 116
   Horizontal transfer 117
Summary and conclusions 122

7. **Renewable energy investment under dominant state ownership: the case of Vietnam** **124**
Introduction 124
Energy supply and renewable energy resources in Vietnam 125
   Overview 125
   Energy trends 127
   Renewable energy resources 128
Structure and liberalization of the electricity supply industry .. 129
   Electricity market and institutions 129
   Privatization and liberalization 132
Renewable energy and technology transfer 135
   Vertical transfer 136
   Horizontal transfer 138
Summary and conclusions 139

8. **Renewable energy investment under strict bureaucracy: the case of Indonesia** **142**
Introduction 142
Energy supply and renewable energy resources in Indonesia ... 143
   Overview 143
   Energy trends 144
   Renewable energy resources 145
Structure and liberalization of the electricity supply industry 147
   Electricity market and institutions 147

    Privatization and liberalization 150
    Renewable energy and technology transfer 152
        Vertical transfer 152
        Horizontal transfer 155
    Summary and conclusions 159

9. Off-grid renewable energy under active investment: the Philippines 163
    Introduction 163
    Energy supply and renewable energy resources in the Philippines 164
        Overview 164
        Energy trends 165
        Renewable energy resources 166
    Structure and liberalization of the electricity supply industry 168
        Electricity market and institutions 168
        Privatization and liberalization 171
    Renewable energy and technology transfer 175
        Vertical transfer 175
        Horizontal transfer 180
    Summary and conclusions 183

Part III Conclusions

10. Renewable energy investment and technology transfer in South-east Asia 189
    Introduction 189
    Investment trends and structures in South-east Asia 190
    Local renewable energy development 195
        Technological choice 195
        Investment and funding structures 196
    Enhancing technology transfer 199
        Vertical transfer 199
        Horizontal transfer 201
    Conclusion and policy recommendations 203

11. Redefining international investment and technology transfer for climate change mitigation 208
Introduction 208
Redefining technology transfer for climate change mitigation 209
    Vertical or horizontal technology transfer? 209
    Northern or Southern technology? 212
    Technology or financial management? 215
Harnessing foreign investment for climate technology transfer 216
    Opportunities for investment 216
    Implications for trade and technology ownership 219
    Flexible mechanisms for climate technology transfer 221
Institutional finance and foreign investment 224
    Direct investment 224
    Portfolio investment 226
Conclusion and policy recommendations 227

12. Enhancing public–private synergy in climate change policy ... 232
Introduction 232
Investment and national technology policy 233
    Foreign investment and national competitiveness 233
    Implications for climate technology 236
Electricity privatization in developing countries 238
    Market and governance structures 239
    Implications for climate change mitigation 242
Building public–private synergy in climate change mitigation 244
    Reducing transaction costs 244
    Responsibilities for public and private sectors 246
Conclusion and policy recommendations 249

Appendix 1 Article 12 of the Kyoto Protocol: the Clean Development Mechanism 252
Appendix 2 Brief summary of renewable energy technologies 254
    References 259

Foreword
by Michael Grubb

The issue of technology transfer between 'developed' and 'developing' countries has bedeviled international negotiations on environment – and other topics – for decades. The Rio 'Agenda 21' agreement stated that access to and transfer of environmentally sound technology should be promoted, 'on favourable terms, including on concessional and preferential terms, as mutually agreed'; a formulation of words that has been incorporated in similar form in the Climate Change Convention and many subsequent agreements.

As Tim Forsyth notes in this book, in practice relatively little has been achieved. He suggests two main reasons: the disjoint between the public-sector debate and the private ownership of most leading environmental technology; and the long-term, complex nature of the process, particularly concerning the costs and commercial risks to private investors. The reality is that private investment now dominates investment, and the concern of this book is to explore how private investment may best be harnessed to help meet both developmental and environmental goals, specifically relating to the energy sector and global climate change concerns.

In the mid 1990s, our Programme conducted research on industrial innovation and environmental policy in the Organization for Economic Coordination and Development (OECD), and the potential implications for 'sustainable industrialization'.[1] We had also conducted extensive work on renewable energy, and on the climate change regime. I was therefore delighted when Japan's New Energy and Industrial Technology Development Organization (NEDO) expressed interest in supporting more in-depth work on the issues surrounding international investment

---

[1] David Wallace, *Environmental Policy and Industrial Innovation,* London: RIIA/Earthscan, 1995; David Wallace, *Sustainable Industrialization,* London: RIIA/Earthscan, 1996.

Foreword

and the transfer of clean energy technologies, particularly renewable energy sources, in the Asian region.

The first stage of that 'joint venture' was a workshop at Chatham House on technology transfer held a few months before the Kyoto conference.[2] A second workshop, in May 1998, focused more specifically on the energy issues in Asia. Both meetings helped to contribute towards the information and thinking in this book, but Tim Forsyth must take all the credit for the structure of the work, selection of case studies and the focus of analysis and conclusions.

To help inform the debate about generalities with specific realities, in this book Tim Forsyth draws lessons from case studies of renewable investment in South-east Asia, based upon visits to these countries in the course of the project. But the book is also well grounded in the broader academic debate about the role of the firm, and of privatization, in national development and international investment. In linking this conceptual understanding with practical experience, and drawing on the specific mechanisms established in the Climate Convention and its Kyoto Protocol, this book constitutes a unique contribution to debates in all these communities.

In the area of climate change, there is a clear mutual interest in enhancing the transfer of advanced technologies. Developing countries want clean, adequate and affordable energy to meet their development needs. Developed countries want them to achieve this without a massive growth in greenhouse gas emissions. How to achieve this remains a topic of considerable confusion, and this book aims to move that debate forward. After it was submitted to the editor, the Buenos Aires follow-up conference to Kyoto agreed to establish a consultative process, with a list of specific questions to be resolved, 'in order to reach agreement on a framework to enhance implementation' of the Convention's provisions on technology transfer. The book could thus hardly be more timely.

This book completes the first phase of a three-year collaborative effort with NEDO. We wish Tim every success in his new post at the Institute

---

[2] Tim Forsyth (ed), *Positive measure for technology transfer under the Climate Change Convention,* London: RIIA, 1998.

of Development Studies. The next phase of work at RIIA will be led by Gill Wilkins, under the guidance of my own successor as Head of Programme, Duncan Brack. It has been a privilege to work with Tim to bring the project this far, and I wish our respective successors every success in building further upon the work presented in this book.

Michael Grubb
Energy and Environmental Programme
December 1998

Preface

This book is the result of research conducted at the Energy and Environmental Programme (EEP) of the Royal Institute of International Affairs (RIIA), Chatham House. Since its establishment, the EEP has developed an international reputation for developing market-oriented and practical ways to implement international environmental policy. This book complements existing work in the EEP concerning climate change policy and privatization and liberalization of electricity markets worldwide, especially *The Kyoto Protocol: a guide and assessment* (M Grubb *et al*); and *Transforming Electricity: the coming generation of change* (W Patterson).

This book would not have been possible without the encouragement and expert assistance of my fellow members of the EEP, and particularly Dean Anderson, Michael Grubb, Walt Patterson and Jonathan Stern. In addition, the collegiality and assistance of Ben Coles, Nikki Kerrigan and Christiaan Vrolijk are also gratefully acknowledged.

Outside RIIA, the research for and accuracy of this book were greatly assisted by Andrew Barnett (University of Sussex Science Policy Research Unit); Francesca Beausang (Cambridge University Judge Institute of Management Studies); Tim Brennand (Shell International); Anthony Fairclough (European Union); Jenniy Gregory and Jonathan Bates (IT Power Ltd.); Gerald Leach (Stockholm Environment Institute); Mark Mansley (The Delphi Group); Damian Miller (Cambridge University Judge Institute of Management Studies); and Louis Turner (Asia-Pacific Technology Network).

Finally, this research was sponsored by the New Energy and Industrial Technology Development Organization (NEDO) of Japan. NEDO is one the world's leading research and development organizations for renewable energy. My thanks go to Yoshiro Echizenya, Yuzo Fuji and William Stumborg.

Acronyms and abbreviations

ACRE	Area Coverage Rural Electrification
AIJ	Activities Implemented Jointly
ANEC	Affiliated Non-Conventional Energy Centre
AO	Administrative Order (of the Philippines government)
ASEAN	Association of South-east Asian Nations
ASTAE	Asia Alternative Energy Unit (of the World Bank)
BOT	Build-Operate-Transfer scheme
BPPT	Baden Pengkaijian dan Penerapan Teknologi (the government agency for renewable energy development, Indonesia)
CASE	Centre for the Application of Solar Energy (of Australia)
CDF	Clean Development Fund
CDM	Clean Development Mechanism
CIDA	Canadian International Development Agency
CKT	Transmission line length (kilometres)
COP	Conference of the Parties (to the UNFCCC)
CORE	Council on Renewable Energy in the Mekong Region
CTI	Climate Technology Initiative
DGEED	Directorate General for Electricity and Energy Development
DOE	Department of Energy (of the Philippines)
DSM	demand side management
DU	distributed utility
EC	electricity cooperative
Egat	Electricity Generating Authority of Thailand
Egco	Electricity Generating Company
EO	Executive Order (of the Philippines government)
ERB	Energy Regulatory Board (of the Philippines)
ESMAP	Energy Sector Management Assistance Programme (of the World Bank)
EST	environmentally sound technology

EVN	Electricity of Vietnam
FAO	Food and Agriculture Organization (of the United Nations)
FDI	foreign direct investment
GCI	Global Carbon Initiative (of the World Bank)
GDP	Gross Domestic Product
GEF	Global Environment Facility
GHG	greenhouse gas
GREEENTIE	Global Remedy for the Environment and Energy Use – Technology Information Exchange (of the IEA)
GWh	Gigawatts per hour
HEP	Hydroelectric power
IBRD	International Bank of Reconstruction and Development (of the World Bank)
IEA	International Energy Agency
IFC	International Finance Corporation (of the World Bank)
IGO	intergovernmental organization
IPP	Independent Power Producer
IPR	intellectual property rights
IRR	Implementing Rules and Regulations
ISO	International Standards Organization
JI	Joint Implementation
JV	joint venture
kV	kilovolts
LNG	liquefied natural gas
LPG	liquefied petroleum gas
MAI	Multilateral Agreement on Investment
mboc	million barrels of oil equivalent
MEA	Metropolitan Electricity Authority (of Bangkok)
MME	Ministry of Mines and Energy (of Indonesia)
MSIP	Municipal Solar Infrastructure Project (of the Philippines)
MT	million tonnes
Mtoe	million tonnes of oil equivalent

MVA	Mega Volt Amps
MW	Megawatts
NEA	National Electrification Administration (of the Philippines)
NEDO	New Energy and Industrial Technology Development Organization (of Japan)
NEPC	National Energy Policy Council (of Thailand)
NEPO	National Energy Policy Office (of Thailand)
NFFO	Non-Fossil Fuel Obligation (of the UK)
NGO	non-governmental organization
NPC	National Power Corporation (of the Philippines)
NTC	National Transmission Company (of the Philippines)
ODA	official development assistance
OECD	Organization for Economic Coordination and Development
OPIC	US Overseas Private Investment Corporation
OSW	ocean, solar and wind renewable energy technology
OTEC	ocean thermal energy conversion
PD	Presidential Decree (of the Philippines government)
PDP	Power Development Plan
PEA	Provincial Electricity Authority (of Thailand)
PEI	Preferred Energy Investments Co.
PLN	Perusahaan Listrik Negara (state electricity utility of Indonesia)
PPA	Power Purchasing Agreement
PSKSK	Pembangkit Skala Kecil Swasta dan Korporasi (the SPP programme of Indonesia)
PURPA	Public Utilities Regulatory and Policies Act (of the USA)
PV	Photovoltaic
QELRO	Quantified Emissions Limitation and Reduction Objective
RA	Republic Act (of the Philippines government)
READ	Renewable Energy Application and Development

Acronyms and abbreviations

REEF	Renewable Energy and Energy Efficiency Fund (of the World Bank IFC)
RENI	Renewable Energy Network of Indonesia
REPSO	Renewable Energy Support Office (of Winrock International)
RESCO	rural energy service company
SBSTA	Subsidiary Body on Scientific and Technical Assistance (to the UNFCCC)
SDC	Solar Development Corporation
SEB	State Electricity Board
SELCO	Solar Electric Light Company (of the USA)
SELF	Solar Electric Light Fund (of the USA)
SHS	Solar Home System
SIG	small island grid
SPP	Small Power Producer
TAP	Technology Assessment Panel (of the UNFCCC)
tC	tonnes of carbon
TEI	Thailand Environment Institute
TERI	Tata Energy Research Institute (of India)
TFAP	Tropical Forestry Action Plan (of the FAO)
UNCSD	United Nations Commission on Sustainable Development
UNCTAD	United Nations Commission on Trade and Development
UNDP	United Nations Development Programme
UNEP	United Nations Environment Programme
UNFCCC	United Nations Framework Convention on Climate Change
USAID	United States Agency for International Development
USIJI	United States Initiative on Joint Implementation
VIR	Vietnam Investment Review
VWU	Vietnamese Women's Union

Executive summary

1. This book is an assessment of the opportunities and impacts of using international private investment as a means to implement international environmental policy. It focuses specifically on the UN Framework Convention on Climate Change (UNFCCC), and particularly the problem of accelerating technology transfer between industrialized and industrializing countries.

2. Technology transfer has been a major headache in the climate change negotiations ever since the UNFCCC was signed in 1992. For many developing countries, the transfer of environmentally sound technology (EST) from North to South was a necessary commitment before they could accept international attempts to implement the Climate Change Convention. However, progess in technology transfer has been slow because much EST is owned by the private sector, which sees technology transfer to potential competitors in the South as commercially damaging and extremely costly.

3. The problem has been aggravated by North–South disputes on the role of foreign investment and climate change policy. Since 1992, most debate on using international investment for climate change mitigation has centred on the terms 'Joint Implementation' (JI) or 'Activities Implemented Jointly' (AIJ) which have proposed to allow industrialized countries to invest abroad as a way of offsetting their own greenhouse gas reduction targets. However, many developing countries have opposed this, considering it to be unfair because it reduces pressure on industrialized countries to reduce their own emissions, and because most debate about JI/AIJ has focused on forestry or carbon sinks, which do not advance industrial technology transfer and are often based on exaggerated claims about their ecological worth.

4. However, the situation changed when the Kyoto Protocol of 1997 created the Clean Development Mechanism (CDM) as a device specifically

Executive summary

for accelerating North–South investment in climate change mitigation. The CDM offers immense potential for enhancing technology transfer via private investment. This book attempts to provide some assistance to the implementation of the CDM by providing discussion of how private investment may be integrated with international technology transfer in this way.

5. The book is divided into three main sections. Part I (Chapters 1–4) discusses the problems of technology transfer, private-sector investment and the climate change negotiations in general. This section outlines the dilemmas of the firm as a unit of economic activity, and the problems that have been experienced in utilizing business activity in environmental policy. In addition, this section outlines dilemmas of rural electrification as one particular example of how international investment may be used for implementing climate change policy.

6. Part II (Chapters 5–9) advances this discussion by examining four case studies of international investment and the transfer of renewable energy technology in Thailand, Vietnam, Indonesia and the Philippines. These sections discuss the relationship between privatization of electricity supply industries, international investment and the adoption of renewable energy technologies, and provide some well-needed examples of how international investment may be used for both domestic and international environmental policies. Conclusions and policy recommendations are made in Part III (Chapters 10–12).

7. The book argues that technology transfer may be classified in terms of 'vertical' transfer – or the point-to-point relocation of technology via foreign investment, and 'horizontal' transfer – or the local embedding and training of local users and manufacturers. Vertical transfer implies that technology remains owned by the investing company, yet horizontal transfer implies transferring ownership eventually to competitors. Discussions about technology transfer under the climate change negotiations to date have generally focused on horizontal technology transfer, and therefore have alienated private investors. Yet according to new theories of economic globalization, vertical technology transfer

might bring many associated benefits of economic growth. Furthermore, for some globally competitive technologies such as photovoltaics (PVs), it may be more effective to harness international producers of technology rather than develop domestic companies that may not be able to compete.

8. This distinction between vertical and horizontal technology transfer was supported by the case studies of South-east Asia. In these countries it was shown that large-scale rural electrification programmes using renewable energy technology were more likely to succeed when high-technology applications were supplied from international investors but then embedded locally with the assistance of specialist development agencies that could provide local expertise. For example, in the Philippines, the Municipal Solar Infrastructure Project uses photovoltaic technology supplied by BP Solar (Australia), yet is supported by a variety of official aid agencies. This programme is more likely to succeed than the so-called 'one million homes rural electrification project' of Indonesia, which is largely dependent on the subsidized use of domestic PV equipment. The case studies revealed that cost recovery for investors was an essential part of the success of international technology transfer, and this depended in part on the successful construction of local mechanisms to provide commercial and technical support for new technology at the local level.

9. The case studies revealed that two key actions were required for accelerating the adoption of renewable energy technology. The first was the need for legislation that provides incentives for private companies to produce their own electricity. This is particularly important for locations where companies have their own power sources. In Thailand for example, the Small Producers Programme allows factories to sell electricity to the national grid, and has been particularly successful with regard to factories that produce biomass waste. In Indonesia, similar legislation has given companies in smaller grid systems the incentive to supply electricity from mini-hydro technology. These examples show that renewable energy development does not have to be in off-grid locations. The second action is the creation or empowerment of specialist

Executive summary xxi

energy agencies that can act as links between international investors and local end users. These organizations, such as Preferred Energy Investments in the Philippines, reduce the transaction costs of investors and work to create local embedding of technology in host countries.

10. There are two key conclusions for the climate change negotiations. The first is that technology *can* be transferred successfully by private investment, but only through redefining the meaning of the term 'technology transfer'. The adoption of climate technologies such as renewable energy may be accelerated in developing countries by allowing such exports to be rewarded by crediting against national greenhouse gas emissions reduction targets. However, this may not imply an increased ownership in climate technologies by developing countries but a greater contribution to other developmental themes such as rural electrification. The CDM may take a lead role in encouraging private investment in climate technologies, and steps should be taken to ensure that investment is used for industrial technology rather than simple reforestation. The establishment of new mechanisms such as international clearinghouses for technology may also assist this process by providing bodies into which both public and private investors may contribute new innovations in return for carbon credits. Other organizations such as the Global Environment Facility or official development assistance might then complement the CDM by assisting with horizontal technology transfer, or the local embedding and training associated with new technologies.

11. The second key conclusion is, however, that the increased incidence of North–South technology flows will have impacts on technology developers in the South. If the CDM effectively subsidizes exports from industrialized countries it will result in an increased adoption of Northern technologies, and a decreased competitiveness of Southern technology producers. These trends may erode the market share currently held by Southern companies and lead to a dependency on imported technologies that are not as appropriate to developing societies as those developed locally. Furthermore, in terms of climate change mitigation, it is also possible that investment from the CDM may only

result in an increased market share for Northern exporters rather than an overall increase in the adoption of renewable energy technologies. If this occurs, then the CDM may only have resulted in the loss of market share by local producers, with no overall reduction in greenhouse gases. The solution to this problem is to increase the ability of the CDM Executive Body and national technology policies to identify which technology development opportunities may best be served through international investment and which through domestic companies.

12. In summary, the conclusion of the book is that private investment is a necessary but not sufficient contribution to international environmental policy. Many dilemmas of technology development and transfer cannot be achieved without producing the appropriate incentives for private companies. However, strong local, national and international interventions are still required to ensure that this private investment may fully achieve the objectives of environment and development policies.

Part I
Themes

Chapter 1

Introduction

International investment and climate change

It is now almost universally acknowledged that international agreements on environmental protection need to be implemented at the level of the commercial company. The reasons for this are clear. Private investment is now rising substantially beyond levels of official development assistance. Government expenditure is shrinking and authorities are increasingly entrusting public-sector policy objectives to the private sector. Multinational firms and international investors now have a global presence that requires an alternative governance system to that developed for the nation state.

But how can firms be involved in international environmental policy? And what are the implications? This book is an attempt to answer these concerns by looking in detail at the interrelationships of foreign investment and climate change mitigation using the case of international investment in renewable energy technology in South-east Asia.

Assessing the role of foreign investment in climate change is particularly topical. Between 1990 and 1997, funding from the Global Environment Facility (GEF) – the body set up to address global environmental problems – amounted to US$5.25 billion. Private-sector transfers from foreign direct investment for the same period reached nearly US$250 billion. Under the Kyoto Protocol of 1997, a new mechanism for harnessing private sector funding, the Clean Development Mechanism (CDM), was created to allow developed countries with obligations to reduce greenhouse gas emissions the ability to achieve these targets by investing in sustainable development projects in developing countries. Foreign investment is increasingly becoming an essential tool of international environmental policy. But relatively little is known about the implications of this trend, or the forms of regulation that may maximize its benefits.

Concurrently with these developments in climate change policy, many countries are also undergoing major new transformations in the way that electricity is generated and supplied to consumers. Privatization and liberalization of electricity supply industries in developing countries is opening up a vast range of opportunities for foreign investment and for decentralized power generation. There is great potential to integrate these changes with international investment in climate change mitigation by increasing the adoption of new renewable climate friendly technologies. In the case of renewable energy technologies, investment may advance climate change policy by reducing dependency on fossil fuels, and accelerate local development by providing electricity without the need for infrastructure such as grid extension or large power stations and dams.

This book analyses how international investment in new energy technologies may be used to advance such environmental and developmental objectives. The book is designed for readers – such as policymakers, investors and students – who wish to know more about implementing global environmental objectives at national and local levels in coordination with the private sector. In particular, the topics of technology transfer and the impacts of privatization are discussed in relation to their role in advancing climate change policy and renewable energy development in developing countries. However, the book also seeks to explain the potential impacts of international investment on both environmental and development objectives.

This initial chapter introduces readers to underlying debates concerning the book's central themes of international investment and climate change policy. Readers experienced in these debates may prefer to move directly to Chapter 2 or to the discussion of the book's outline.

Environmental policy and renewable energy development

The focus of this book on climate change policy might seem a curious starting point for an analysis of rural electrification and renewable energy development. Indeed, decentralized energy development is often justified on the grounds of poverty alleviation and economic growth, rather than national or global environmental concerns (see Barnett *et al*, 1982; Gregory *et al*, 1997; IEA 1997 a and b; Kozloff, 1994; Ramani *et al*, 1993).

This book does not contradict this concern with local energy development and poverty alleviation but instead aims to complement these debates by assessing the impacts of international investment and privatization on energy development. This focus does not place greater importance on international environmental policy over local development, but assesses the pros and cons of such new policy. As international agreements on environmental policy become stronger and more numerous, it is important to assess both how policy can be implemented at the local scale, and also the impact policy has on local regions or countries. The aim of this is to ensure that policy can achieve both international environmental and local development benefits.

The concern about the potential negative impacts of international policy is well justified. Recent studies have indicated that many so-called 'global' environmental problems, such as climate change, have varying significance in different countries in terms of physical impacts such as rising sea levels, changing agricultural productivity and altered vectors for diseases like malaria. Yet in addition, social scientific studies have questioned to what extent 'global' problems are given equal weighting in political terms by different countries, or how proposed solutions to problems may be acceptable to all countries (Buttel and Taylor, 1994; Rayner and Malone, 1998; Yearley, 1996). As discussed in subsection (c) below, some proposals for climate change mitigation have involved reforestation or investment in developing countries that may be used to offset national obligations for abating greenhouse gas emissions, but which have either been opposed by developing countries, or have controversial local impacts. The ability to identify solutions to climate change that are acceptable to both North and South is therefore of significance to both political acceptance of climate change policy and local development.

Investment in renewable energy is one potential action for climate change policy that may advance both local development and international environmental policy. However, it is acknowledged that other energy technologies – such as natural gas pipelines, clean coal technology and energy efficiency technology – may have a greater and more immediate impact on climate change mitigation (see Grubb and Walker, 1992; Mitchell, 1997; Paik, 1995). These other technologies are not discussed in this book, which instead focuses on renewable energy as an example of one forward-looking form of climate technology that may provide benefits to both North and South. Current contributions of renewable energy to national energy consumption are extremely small in comparison with fossil fuels, yet the potential for expansion is huge. Grubb *et al* (1997), for example, estimate that new renewable energy technologies may account for some 25 to 50 per cent of European electricity supply by 2030. In the developing world, where there are allegedly two billion people without access to grid-supplied electricity (World Bank, 1996), the potential for decentralized electricity production through renewable energy is vast. This book may therefore help accelerate the development of renewable energy in developing countries by assessing the impacts of international investment and environmental policy.

Technology investment and privatization

Another key debate addressed by this book is the role of privatization and international investment in the development or adoption of renewable energy technology. In the UK and the United States, much success was made in accelerating private investment in renewable energies respectively by the Non-Fossil Fuel Obligation (NFFO) and Public Utilities Regulatory and Policies Act (PURPA) (see Grubb *et al*, 1997). These legislative packages were introduced during the 1980s at the same time as electricity generation and distribution were undergoing privatization and liberalization, and stipulated that utilities should buy a small proportion of electricity from alternative energy sources. As a result, there were rapid advances in photovoltaic (PV), wind turbine and biomass generators, which in turn helped reduce the previously high costs associated with these technologies.

The success of these measures suggests that similar programmes could also build renewable energy investment in developing countries. However, there are some important differences between economic reforms occurring in North and South. The term 'privatization' has been used generally to indicate a growing dependency on private-sector suppliers to public-sector infrastructure or policy objectives (see Clarke and Pitelis, 1993). Yet the individual reforms leading to privatization may differ between countries, and include various associated actions in addition to increasing private ownership. Such additional reforms may include commercialization, or 'corporatization' of state-operated enterprises to make them act under strict business conditions; the 'unbundling' of state monopolies into several smaller entities; and the increase in wholesale or retail competition between suppliers to add efficiency (USAID, 1998).

Privatization in industrialized countries such as the UK and the United States has normally been associated with the deregulation of state monopolies, or the sale of existing infrastructure such as the national grid. Furthermore, another aim of privatization and liberalization of generating authorities has been to increase efficiency by introducing competition between different suppliers. Reforms have also taken place against the backdrop of a mature electricity market, where consumers are used to billing structures, and where regulatory agencies are comparatively well developed.

In developing countries, however, privatization has generally been introduced as a way to accelerate the funding of public infrastructure projects that have not yet been constructed. In addition, liberalization has often meant the separation of the State Electricity Board (SEB) from government into a separate entity without an increase in competition. As a consequence, privatization in many developing countries is not associated with deregulation or some of the impacts of liberalization in industrialized countries. In countries where there are low rates of rural electricity supply, it may still be necessary to maintain cross-subsidization or deliberate market intervention in order to increase investment in rural electrification, or other social aspects of electricity policy. The process of privatization in developing countries, therefore, may need the parallel development of systems of regulation in order to ensure that increased private investment can still achieve public policy objectives outside the provision of power alone (Berg, 1997; Bruggink, 1997; USAID, 1998).

A further topic for concern is the impact of privatization and foreign investment on the ownership of technology. In the UK, the introduction of the NFFO accelerated investment in renewable energy technology, but also led to an influx of Danish wind turbines that greatly damaged the UK indigenous wind turbine industry. Increasing international investment in renewable energy technologies in developing countries may have a similar effect in displacing indigenous technologies, and increasing dependency on foreign imports or locally manufactured, but foreign owned goods.

According to traditional approaches to economic and technological development, such trends may be seen to be damaging to local industries, and also the ability of host countries to compete in international markets (see Porter, 1990). However, these traditional approaches are being increasingly challenged by arguments that take into account the growing globalization of technology investment and markets (see Reich, 1991; Howells and Michie, 1997). Under these newer approaches, international investment and ownership of technology may bring many benefits to regions because of employment, training, regional specialization and the attraction of further investment. International investment may also be necessary to gain access to certain sophisticated technologies that are globally competitive and manufactured abroad by multinational companies, rather than by attempting to develop these technologies through a long-term programme of indigenous industry development (Cantwell, 1989; Dunning, 1997).

International investment may therefore accelerate local development and lead to the adoption of certain technologies. However, it may also expose countries to international competition that may damage indigenous industries, and increase dependency on foreign technologies. Privatization and international investment may further encourage vertical integration of companies, in which firms merge or open new subsidiaries in order to minimize costs by keeping all manufacturing and distribution activities within one large company. These actions decrease the profits available from investment to local companies and societies, and also restrict the extent to which international investment may contribute to technology transfer.

Such impacts have particular relevance for renewable energy investment in developing countries. Many renewable technologies such as PV and wind turbines are well developed in industrialized countries, and could accelerate decentralized rural electrification in developing countries. However, there are also many renewable technologies in the South that may also achieve this purpose. In particular, India is a world leader in the construction of some biomass-using generators, and there are also many examples of PV or wind turbines in other developing countries that are serviceable if not competitive with Northern imports (World Energy Council, 1994; TERI, 1997).

The encouragement of international investment as a way to accelerate renewable energy development and dissemination is therefore politically sensitive, and has to be integrated with strong national industrial and technological development policies to identify which technologies may be developed indigenously, or through foreign investment. This book addresses these concerns by discussing the implications of using international investment for renewable energy development, and then illustrating these in four countries of South-east Asia to show how foreign investment has impacted on local energy development.

Climate change policy and technology transfer

Finally, this book addresses some important concerns within the climate change negotiations relating to foreign investment and technology transfer. Ever since the UNFCCC was signed in 1992, the topics of foreign investment and technology transfer have been bitterly contested between North and South, and are representative of many of the divisions that exist between them.

The transfer of environmentally sound technology (EST) between North and South is an essential requirement in order to enable countries to undergo rapid industrialization without greatly increasing greenhouse gas emissions. In addition, many developing countries have made it a political condition of signing the UNFCCC and related international agreements (see Grubb *et al*, 1993). However, the term is hotly debated, and developing countries have complained that comparatively little has been achieved since 1992 (Forsyth, 1998).

The lack of progress in international climate technology transfer is often explained by two key factors. Firstly, international agencies have pointed to the growing importance of private investment in North–South financial flows and the difficulty of harnessing this investment for public-sector environmental policy initiatives. Secondly, technology transfer itself is a long-term complex process involving costs and commercial risks to private investors. Indeed, the term 'technology transfer' is invariably used only in public policy discourse. Private companies use terms such as 'joint ventures', 'contracting' and 'licensing' as ways to denote the commercial basis for technology usage, and have little interest in achieving what is generally known as technology transfer (Heaton *et al*, 1994; MacDonald, 1992). As a result of these two factors, technology transfer has largely failed because of the inability to offer adequate incentives to private-sector actors.

Such incentives may come from the adoption of so-called 'flexible mechanisms' of climate change mitigation, or those that employ market-based incentives or international investment to gain adherence to the UNFCCC. Before the Kyoto Protocol, most debate about flexible mechanisms referred to emissions trading and Joint Implementation (JI). Emissions trading allows Annex I countries to achieve emission commitments[1] by trading permits to emit between each other. JI allows Annex I countries to offset commitments by investing in foreign countries in projects that mitigate climate change.

Flexible mechanisms were bitterly opposed by many developing countries before the Kyoto Protocol because they were seen to be ways for Annex I countries to avoid taking responsibility for action at home.

---

[1] Emission commitments are commonly referred to by the acronym: QELROs – Quantified Emissions Limitation or Reduction Objectives.

In addition, mechanisms are difficult to measure and easy to manipulate. In the case of emissions trading, generous emissions commitments awarded to countries that have undergone industrial decline such as Russia and the Ukraine may lead to the so-called 'hot air problem'. 'Hot air' would be created if countries such as Japan and the USA simply buy their allocations from these depressed countries, resulting in no overall reduction in greenhouse gas emissions. Similarly, JI has been criticized for being difficult to assess over what might have occurred anyway (the 'baselines problem'); for allowing international investors to select only the cheapest options and locations (the 'cherry-picking problem'); and because JI (and its associated mechanism, Activities Implemented Jointly – AIJ) has focused attention on carbon sinks or forestry projects rather than projects that transfer industrial technology.

Under the Kyoto Protocol, however, three flexible mechanisms were created. Emissions trading and JI were finally agreed for countries within Annex I. However, a new flexible mechanism – the CDM – was created to address investments between North and South. The CDM is similar to JI in that it allows Annex I countries to offset part of their commitments by investing in climate change mitigation in foreign countries. However, the CDM is different to JI by dealing specifically with non-Annex I (usually developing) countries, by establishing it directly under the supervision of a multilateral Executive Board, and by a subtle difference in the types of projects to be advanced. In particular, the text of the Kyoto Protocol referring to the CDM (Article 12, see Appendix 1), states that investment should be used for 'sustainable development' projects in non-Annex I countries, and the word 'sinks' is not mentioned once. Debate continues as to whether forest projects should be included under the CDM.

The potential therefore exists to use the CDM to advance international investment in EST or renewable energy technology that may have the double benefits of climate change mitigation and local economic development. However, the structures and incentives necessary to achieve these aims are still not clear. As discussed above, international investment may accelerate the adoption and local manufacture of new renewable energy technologies, but they may also damage the competitive standing of existing Southern technologies that may be more appropriate. In addition, it is not clear how such direct point-to-point investment in new technologies may satisfy Southern demands for technology transfer under the UNFCCC.

This book addresses these concerns by examining critically the concept of technology transfer as adopted under the UNFCCC and relating it to new approaches to technology development under the globalization of international investment. Four case studies of countries in Southeast Asia are then assessed to identify how far international investment may lead to technology transfer and the successful development of new renewable energy electricity sources. It is argued this information answers many important questions concerning how to make international investment for climate change policy effective for both international environmental policy and local development.

Aims and structure of the book

The key questions addressed by this book are:

- What are the implications of harnessing international investment for climate change mitigation?
- What incentives and forms of business can allow maximum transfer of technology to developing countries via foreign investment?
- How may the privatization and liberalization of electricity supply industries in developing countries be managed to ensure maximum adoption of renewable energy technology?
- What national and industrial policies are necessary to ensure an acceleration of international investment in renewable energy but without harmful impacts on indigenous industries or local end users of technology?
- How may the uses of international investment and technology be redefined in the climate change negotiations in order to overcome serious political divides between North and South?

Introduction 13

This book is deliberately different from many other books on renewable energy and technology transfer. It does not seek to identify the technological constraints of new renewable energy systems (for example, see Ahmed, 1994; Green, 1997). The book also does not attempt to model or measure the historic progress of transferring technology to developing countries via patents and licenses (for example, see Ohkawa and Otsuka, 1994). Part of the reason for this is because foreign investment in renewable energy investment in Asia is comparatively new and some forms of renewable energy industries such as PV are still poorly developed.

Instead, this book aims to assess which institutional and business structures are most efficient for maximizing the impacts of foreign investment on renewable energy development and technology transfer. It is believed that an in-depth analysis of technology transfer, international investment and renewable energy technology in developing countries will assist the climate change negotiations by attempting to unlock North–South divides and recognize practical mechanisms to accelerate private investment in climate change mitigation.

The region chosen for analysis is South-east Asia. South-east Asia underwent financial crisis in 1997–8 as a result of poor financial management and a decline in competitiveness for exports. This crisis does not undermine the purpose of the book, but instead makes it more relevant. The governments of South-east Asia have reiterated that they intend to proceed with energy development despite the short-term impacts of the crisis, and so still seek incentives to attract international investment for public policy objectives. In addition, the ability of renewable energy investment to lead to decentralized rural electrification without the need for large-scale infrastructure such as grid extension or dams may also be attractive for reducing the costs of public policy objectives.

Furthermore, South-east Asia offers a variety of countries with different regulatory regimes for investment and physical conditions for renewable energy development. Thailand, Vietnam, Indonesia and the Philippines have been selected as case studies in order to allow comparisons to be drawn and lessons to be identified that may be applied elsewhere.

There are three parts to the book:

- Part I identifies the key themes of the debates concerning international investment and climate change. These chapters detail the theoretical and historical elements to the climate change negotiations, technology transfer and renewable energy, and privatization and energy investment in South-east Asia.
- Part II presents the main empirical work of the book by analysing in detail case studies from Thailand, Vietnam, Indonesia and the Philippines.
- Part III concludes by presenting three chapters that assess the lessons to be learned from the case studies. The first assesses lessons for the local debate of building renewable energy and technology transfer in Asia. The second identifies the significance of the information and analysis for the climate change negotiations. The final chapter discusses which forms of business–government interrelations may achieve effective public–private partnerships in climate change mitigation.

The book's intended readership includes policymakers, investors and specialists outside academia who are concerned with integrating foreign investment and climate change mitigation. Students of development or environmental policy may also profit from the analysis of incorporating privatization into public policy objectives.

The book adopts a theoretical approach broadly inspired by institutional economics, or the study of firms as an active unit of economic activity alongside free markets and government regulations (see Tidd *et al*, 1997; Williamson, 1975, 1985). The aim of this approach is to identify those combination of market forces, government restrictions and company activity that may best lead to the achievement of public policy aims. One key concept of institutional economics is transaction costs, or the costs of interacting with other firms and organizations via contracts or other arms-length agreements. Reducing transaction costs facilitates economic efficiency and increases the ability to utilize private-sector investment in public-sector policy. The book also focuses on the formal institutions of environmental and industrial policy, such as firms, legislation and international development agencies. It is acknowledged that other, informal institutions such as networks between investors and policymakers may also be relevant (Harriss *et al*, 1995; O'Riordan and Jordan, 1996).

The next chapter builds on this introduction by discussing in more details the dilemmas facing policymakers of incorporating private investment in climate change policy. The key argument is that privatization and internationalization of environmental policy includes many political and economic implications that are not always recognized at the time policies are proposed.

Chapter 2

International investment and climate change mitigation

Introduction

This chapter provides a discussion of the main theoretical themes of the book concerning the integration of private-sector investment into climate change policy. There has been much discussion about harnessing foreign investment for climate change mitigation, but little attention has been paid to which incentives or regulations are necessary to achieve this integration, or the potential political and economic impacts. This chapter addresses these concerns by assessing current debates about using the firm as a tool of environmental policy, and how this may shape global climate change policy.

The chapter is divided into three main sections. The first section assesses the potential of harnessing foreign investment for climate change mitigation with particular reference to privatization and liberalization of electricity supply industries. The second section then analyses the economic and political implications of utilizing the firm in environmental policy and the successful achievement of public–private synergy. The third section then considers the role of foreign investment in climate change policy, with particular attention to JI and technology transfer. It is acknowledged that environmental policy is not the only driving force for renewable energy development in developing countries.

The key argument of this chapter is that private investment presents a wide range of benefits for environmental policy and climate change mitigation. However, privatization and the harnessing of international investment may also have political implications concerning the ownership of technology, national competitive standing and the selection of priorities for environmental policy. These problems may also be marked when investment from industrialized countries is undertaken in developing countries. It is therefore argued that harnessing private investment in environmental policy should not be seen as simply trusting in market forces, but instead as a redistribution of responsibilities between public and private sectors, which in turn requires continued governance of both the investment process and its environmental impacts.

The attractions of harnessing foreign investment

The growth of the private sector

The need to incorporate private investment effectively in international agreements on environment has never been greater. In virtually all countries of the world, the ability of states to invest public-sector funds in environment or development projects is diminishing, but the strength and presence of the private sector is increasing. Statistics make this clear. In 1993, foreign direct investment (FDI) from North to South amounted to US$64 billion, while official development assistance (ODA) had stagnated at US$59 billion. If this figure is adjusted to include portfolio investment, which stood at US$87 billion in 1993, private capital flows amounted to three times the size of ODA. In 1996 private flows were US$244 billion: four times the size of ODA (World Bank figures, in Chung, 1998:48).

Foreign investment is now the greatest financial flow between developed and developing countries, and this represents increasing environmental impacts of industrialization, and also the ability to use this investment for environmental and development purposes. Tables 2.1 and 2.2 show the growth and importance of foreign investment between North and South, although much of this investment currently targets the more developed countries of Asia and Latin America, rather than Africa. The indirect impacts of portfolio investment – in funds and stock – also need consideration in addition to FDI.

In addition to the growth of private investment, privatization – or the increasing private-sector supervision and achievement of public-sector objectives – is increasing as a policy framework throughout the world. The attraction of privatization is that it allows the achievement of public-sector goals with supposedly greater efficiency and accountability, and with the potential to use international environmental or technological standards and expertise. In Europe and North America,

Table 2.1: International private capital flows to the top 12 developing countries,[1] 1996[2]

Country	Total flows (US$ bn)	Share of GDP[3] (%)	Amount per capita (US$)
China	52	7	42
Mexico	28	5	294
Brazil	15	3	89
Malaysia	16	14	777
Indonesia	18	6	89
Thailand	13	5	224
Argentina	11	3	323
India	8	1	8
Russia	4	0	25
Turkey	5	1	74
Chile	5	6	317
Hungary	3	18	248

Source: French (1998:10)

Table 2.2: International private capital flows to developing countries, 1990 and 1996

Source	1990 Amount (US$ bn)	1990 Share (%)	1996[4] Amount (US$ bn)	1996[4] Share (%)
FDI	25	56	110	45
Portfolio equity (stocks)	3	7	46	19
Portfolio debt (bonds)	2	5	46	19
Commercial bank loans	3	7	34	14
Other[5]	11	25	8	3
Total	44	100	244	100

Source: French (1998:11)

[1] The top 12 developing countries are ranked according to those that received most cumulative private capital inflows from 1990 to 1995 (using World Bank figures)
[2] 1996 numbers for private flows are preliminary
[3] 1995 numbers
[4] Preliminary numbers from World Bank
[5] Principally export credits from companies and official export credit agencies

privatization has been associated with liberalization, or a process of decentralizing power from state bureaucracies and introducing competition to state-operated monopolies. However, it is important to note that the nature of privatization varies between countries. In many developing countries, privatization has normally been introduced as a way to accelerate funding for public infrastructure projects rather than a process of liberalization or deregulation. Indeed, in some countries where electricity systems are poorly developed, private investment may be associated with providing the infrastructure for new energy monopolies that are still controlled by the state. Alternatively, privatization and liberalization may imply the separation of SEBs from governments into separate entities without increasing overall competition. Furthermore, there is also concern from industrialized countries about how far privatization has improved the achievement of public-sector objectives, or merely reduced the costs associated in doing so (Berg, 1997; Bruggink, 1997; Ripple and Takahoshi, 1997).

Private investment, and the process of privatization in general, therefore have immense potential in shaping the size and nature of investment. However, the introduction of privatization alone does not always increase competition, or lead to a change or improvement in public policy since the time when investment was state dominated. There is consequently a need to identify ways in which private investment may be used effectively for public policy objectives. Commonly, this may mean the development of an independent regulator or governance mechanism that allows privatization to be accompanied by a re-evaluation of policy objectives and full monitoring of new investment.

Privatization and liberalization of electricity supply industries

The trend to privatization is particularly apparent in the electricity supply industries. The established mechanisms of electricity generation, transmission and distribution to consumers is on the brink of immense changes as state ownership and responsibility for supply is being replaced by private investors (see Patterson, 1999). The transition has major implications for the options available to consumers, the responsibility of business operators and the ability to integrate electricity development with environmental policy.

Firstly, the privatization and liberalization of electricity impacts on the economic efficiency and nature of electricity businesses, Under the established approach of state ownership, electricity generation, transmission and distribution were all undertaken by one central body – the SEB – and this encouraged the centralization of production via a few large power stations, and the transmission via one large national grid system. The liberalization and privatization of these functions allows a variety of new technologies and options for supply. In particular, reforms may lead to the decentralization of electricity generation and supply, possibly without connection to a grid. It also gives local energy companies a greater ability to diversify into electricity generation, as has been seen in the UK with the production of power by British Gas using offshore gas fields.

Secondly, the responsibility for electricity supply is turning increasingly to private-sector investors. Under state ownership, the government held the responsibility for ensuring power supply as part of its assumed duties. Under privatized ownership, however, this assumption may change according to who has the contract to supply electricity. This was illustrated in 1998 when an ice storm in Canada cut off power supplies to several towns in southern Ontario. It was commonly assumed that the storm had damaged all equipment, and that the reduction in supply was an inevitable hardship for all consumers. In fact, the storm had only damaged some of the local electricity distribution infrastructure, and other plants would still have been able to supply electricity had there been a commercial contract in place. The privatization of electricity supply, therefore, may be associated with a reduction in costs for consumers, but may also lead to a reduction in the security of electricity supply.

Thirdly, electricity privatization and liberalization are of immense significance for energy and environmental policies. In the UK, the transition to private ownership since the 1980s has been associated with the increased use of cogeneration and natural gas as a source of fuel. The old Central Electricity Generating Board had no immediate commercial incentives for adopting either of these practices, and the centralized nature of planning within the organization tended to make it bureaucratic and resistant to change. The liberalization process can therefore accelerate the adoption of new technology and ideas. Furthermore, the growth of a decentralized electricity supply increases the potential for using new and renewable sources of energy on either an off-grid or grid-connected basis.

There is consequently great potential for integrating the privatization and liberalization of electricity supply industries with environmental measures to reduce greenhouse gas emissions by adopting alternative fuel sources and energy-saving technology. However, integrating different policies in this way requires an understanding of the mechanisms and implications of private investment.

Utilizing the firm in public and environmental policy

The preceding section established that the private sector is of growing importance in international investment, and that privatization and liberalization of electricity may bring immense opportunities to environmental policy. But how can the private sector be incorporated into environmental policy? This section addresses this question by describing the economic internalization process that creates firms as independent economic units, and the implications of this process for public policy.

This section has four subdivisions. The first describes the nature of firms as economic units, and the process of vertical integration that may lead to these. The second subdivision discusses how privatization and newly integrated firms may contribute to public policy objectives. The third concerns the implications of global private investment on technological development. The fourth considers the implications of private-sector involvement in environmental policy concerning the identification of environmental objectives and the fulfilment of policy aims. It is argued these discussions are crucial to understanding both the pros and cons of harnessing private investment in international environmental policy.

The firm as an economic unit

A common perception is that a firm is a company or business trading under one name. However, in theoretical terms a firm is a complex economic unit and this needs to be understood before it is utilized in public or environmental policy. The economic definition of a firm is the internalization of economic activities within a single trading unit. A firm is drawn into existence when it is more profitable to undertake activities within that unit than to conduct costly contractual procedures with other trading units (Coase, 1937). The costs of dealing with other firms or organizations in the market place are known as transaction costs, and these need to be minimized in order to increase the efficiency of economic activities. The academic discipline of institutional economics is the study of transaction costs and the formation of firms in order to avoid these (see Williamson, 1975, 1985; Pitelis, 1990).

A key factor underlying the creation of firms is internalization, or the inclusion of economic activities within itself in order to save costs. An important part of this process is vertical integration.

> Vertical integration can be defined as the combination of technologically separable and sequentially related economic activities within the confines of a single firm. Vertical integration occurs when one firm merges either with a firm from which it purchases the inputs or with a firm to which it sells its output. (Wu, 1992:5)

Multinational companies are therefore highly vertically integrated enterprises. The potential advantages of vertical integration are that it can reduce transaction costs and so increase efficiency, and also increase the geographical spread of manufacturing and marketing operations. In particular, vertical integration can reduce costly duplication of research and development or administration between two firms. In addition, integration may also result in the regularization of manufacturing, employment and environmental standards within a firm's activities.

However, the term vertical integration is debatable and can take various forms. Vertical quasi-integration, for example, refers to gaining the advantage of vertical integration but without the full integration of ownership. Also, firms may occasionally form alliances or joint ventures in order to gain market entry or necessary expertise for projects when they have similar aims. When these activities are governed by a contract they are not strictly speaking forms of integration. However, they may be the best options available to companies in investment regimes, such as in many developing countries, where simple vertical integration or direct market entry is not allowed because of government policy aimed at protecting indigenous industries from competition.

The creation of firms, and the role of management in deciding long-term strategic direction, therefore, shows that economic development is not simply controlled by markets (the neo-classical model) or governments (the hierarchical, or statist model) but by intermediary economic units with the ability to shape local markets (Penrose, 1959). Shrewd judgement about when and how to integrate assists firms in overcoming the bounded rationality (or short-term perspectives) of their own organization, and also avoids opportunistic competition from other firms. Environmental policy, therefore, needs to integrate this understanding of company motives into strategies that intend to influence company behaviour.

Firms then, are economic units which are drawn into existence by the forces of competition and the transaction costs of doing business. They can shape economic development by their activities, but changes in industrial regulation and competitive conditions can also shape firms. Vertical integration is the key process by which firms are created, but integration may also lead to monopolization, which can have associated negative impacts such as market exploitation and a decline in innovation and technological development (see Lee, 1994). The management of these problems is now discussed.

Privatization and public–private synergy

The power of firms in the economy is increased by the process of privatization. Privatization is now a global trend, although the economic and political implications of the process are still poorly understood.

Privatization as a framework for policy emerged during the 1980s after reforms, particularly in the UK, of state bureaucracies and state-operated enterprises. The theoretical motivations for privatization are the fear that communal ownership of resources may lead to a 'tragedy of the commons'; the belief that open competition may increase efficiency and fertility of ideas; and the commercialization incentive in order to justify bureaucracies to taxpayers. However, there were other political priorities associated with privatization in the 1980s including the need to reduce government expenditure; easing problems of public-sector pay awards by weakening trade unions; and widening share ownership (Clarke and Pitelis, 1993).

It was therefore hoped that privatization would achieve public–private synergy, or the accomplishment of public-sector policy objectives through the activities of investment, to the mutual advantage of both public and private parties. Public–private synergy may also be associated with the gradual replacement of command-and-control forms of regulation based on negotiation between government and private firms, with the intention of making regulation serve and empower industrial performance rather than restrict it. However, privatization has not meant, and cannot imply, a total removal of any market intervention by government. Instead, privatization and liberalization imply the reorganization of regulatory structures and incentives in order to facilitate private ownership and the achievement of public-identified policy objectives. One classic topic of debate is the role of so-called 'natural monopolies', such as the national grid or nuclear power generation, which may not be best served by distributing ownership. Furthermore, some privatized industries may eventually become monopolized through integration, which may lead, ironically, to less efficient or more exploitative firms.

The political effects of monopolization and privatization are highly controversial and still unclear. Traditionally, debate has varied between analysts who argue multinational companies exploit market positions for power and collusion (eg Hymer, 1976), and those who see internalization and the fear of competition as a way to improve economic efficiency (eg Williamson, 1985). An optimistic middle ground in this debate is the belief that internalization can enhance both innovation and competitive standing by allowing firms to increase technological ability (Cantwell, 1989; Dunning, 1994). The impacts of privatization may vary, therefore, according to the nature and interactions of firms, and the ability of governments to regulate the exploitative aspects of monopolization through national industrial policies. Yet for some countries such national policies or regulatory bodies do not exist, and so privatization might be occurring in locations where it may not be regulated, or where the objectives of public policy are not enforced.

> Privatization works best where markets are lively, where information is abundant, where decisions are not irretrievable, and where externalities are limited. It works worst where externalities and monopolies are abundant, where competition is limited, and where efficiency is not the main public interest. (Kettl, 1993:39)

Privatization is often discussed as an overall direction of public policy implementation, but the incorporation of objectives other than simple economic efficiency may easily be overlooked. However, privatization may be the best way to enhance business investment in selected industries by reducing the transaction costs of market entry and contracting. Some governance of the activities of multinational companies has been achieved via multilateral conventions such as the 1966 Andean Investment Code, or the 1971 OECD International Chamber of Commerce Guidelines for International Investment (Clarke, 1983). But guidelines vary at the national level, and may be absent for some recently privatizing industries such as the electricity supply industries in developing countries.

Some general principles concerning the governance of investment during privatization are already clear. Box 2.1 shows a list of recommendations from one analyst for enhancing public–private synergy through stringent and cooperative policy formulation with business.

Global investment and technological competitiveness

The power of firms is also increasing internationally as a result of multinational company activities, growing private sectors in developing

> **Box 2.1: The smart-buying bureaucracy: suggestions for maximizing public–private synergy during liberalization and privatization**
>
> - Shrinking the state does not mean losing control – smaller bureaucracies need to be smarter bureaucracies
> - Maintain quality staff – hire and reward bureaucrats trained to manage contracts, and retrain mid-level bureaucrats
> - Improve government–business relations – make political appointees aware of the issues included in contracting
> - Keep privatization in perspective – tone down political rhetoric about private-sector involvement
> - Don't lose sight of public policy objectives – avoid contracting for core government functions
> - Maintain control on processes – recognize that market methods raise new issues for governance
>
> Source: Kettl (1993:208-10)

countries and the adoption of privatization as a framework for policy in developing countries. Privatization is encouraged in many developing countries because it promises greater economic efficiency and enactment of public policy objectives at a time when few states have the ability to spend large amounts on infrastructure or public services. Private investment also allows governments to harness new technologies from foreign investors, either for service provision or for upgrading existing technology in indigenous firms (UNCTAD, 1997; World Bank, 1995).

However, privatization in the developing world is often different to privatization in countries such as the UK and the United States. Privatization in Europe and North America has often been motivated to increase efficiency of existing state-operated enterprises. In industrializing countries privatization is often sought to provide funds and expertise for infrastructure and organizations that do not yet exist. Consequently, private investment in developing countries through Build-Operate-Transfer (BOT) schemes and their associated forms may be to supply infrastructure for non-liberalized state enterprises, or for monopolistic structures such as state electric utilities. Indeed, it is not clear what political impacts of privatization may occur in locations where regulatory structures and even industry itself are poorly developed.

In the developing world, whilst economic change is essential, the imposition of privatization without a fundamental renegotiation of the terms of trade and aid with the developed world is likely to achieve little more than the acute disappointments of central planning. (Clarke and Pitelis, 1993:26)

Increasing private and foreign ownership means a number of choices and dilemmas for the governments of developing countries. On one hand, they offer new investment and technology; on the other they often imply a greater dependency on imported technology and a loss of indigenous competitiveness. The ideal, yet elusive, strategy is to find a system of governing private foreign investment that allows indigenous industries to grow while also allowing foreign investment to support the economy. Such national industrial policies can have great success, and indeed some have argued that the growth of the East Asian 'tigers' of Taiwan, Korea, Singapore and Hong Kong between 1960 and 1990 was a result of careful national industrial policies that allowed foreign investment to produce exports while protecting indigenous infant industries (Cantwell, 1989; D. Forsyth, 1990; Stoneman, 1987; Wade, 1990). These arguments contradict alternative explanations that have proposed that free market forces allowed East Asia to prosper, and indeed such national industrial policies may have even greater significance after the 1997–8 financial crisis.

The debate is also relevant for national competitiveness and technological development policy. A well-established belief is that a nation's competitiveness in global markets depends on the success of its firms (see Porter, 1990). However, new approaches are increasingly suggesting that it is futile to attempt to build indigenous industries in globally competitive high-cost industries, and that governments should instead seek regional development by inviting international firms to invest within national boundaries (Howells and Michie, 1997; Lall, 1996; ul Haque, 1998).

Under such an international approach, firms may wish to invest in regions according to the regional specializations in resources, labour or regulatory environments. However, the implication of this approach is that technological ownership remains in foreign hands, and this may have an impact on which technology is developed (so-called path dependency of technology) and the loss of competitiveness of indigenous technology producers. Yet this kind of investment also brings other benefits, such as employment, training and multiplier effects on local expenditure that may have a greater impact on local development than long-term strategies to develop technological expertise in indigenous industries. Indeed, according to economic theorist Robert Reich (1991:264):

> A work force that is knowledgeable and skilled at doing complex things, and which can easily transport the fruits of its labour into the global economy, will entice global money to it. The enticement can develop into a virtuous relationship: well trained workers and modern infrastructure attract global webs of enterprise which invest and give workers relatively good jobs; these jobs, in turn, generate additional on-the-job training and experience, thus creating a powerful line to other global webs. As skills increase and experience accumulates, a nation's citizens add greater value to the world economy – commanding even higher compensation and improving their standard of living.

The impacts of international investment for national economic and technological development are therefore controversial and subject to constant review. Under increasing globalization of investment and technological expertise, the benefits of international private investment for technology development may be immense. However, investment needs to be evaluated for its potential contribution and integrated into national economic policies that seek to maximize its benefits for each country or region. Again, the introduction of private investment and freer market forces has great potential for increasing the achievement of local development or public policy objectives. But these changes have to be monitored and shaped by local incentives and regulations to avoid experiencing disadvantages.

Business investment and environmental policy
This subsection assesses the significance of these debates about public–private synergy and foreign investment for climate change policy.

This in part requires an initial discussion of the pros and cons of integrating business institutions into environmental policy in general. In particular, the inclusion of business into environmental policy represents a paradox because many environmental problems such as waste and pollution result from their actions and addressing these may raise costs, and hence be resisted by business. Including business in environmental policy therefore represents a complex blend of opportunities and problems.

The opportunities of privatization and foreign investment for environmental management and policy refer to the ability to use investment for environmental purposes, and the enforcement of international environmental standards through vertical integration and sales. For example two analysts wrote:

> Preliminary analysis also suggests that privatization (a major source of FDI in many countries) yields significant environmental benefits. Privatized companies generally attract better management, which in turn results in reduced waste and lower pollution. Compared with public owners, foreign investors usually insist on greater efficiency in the operations in which they invest... In general, investment flows from the North to the South seem likely to produce fewer environmental problems than those from the South to other countries in the South. (Esty and Gentry, 1997:162–6)

In particular, analysts have given attention to so-called 'no-regrets' or 'win-win' strategies. These are actions that have environmental benefits, yet are also in the commercial interest of companies. No-regrets actions may include cost-cutting measures such as recycling waste, or value-adding actions such as 'green marketing', which appeal to consumers as an example of a firm's or good's environmental performance. It is therefore increasingly possible to be both 'green' and competitive (Porter and van der Linde, 1995). In addition, the potential negative impacts of environmental legislation on firms' economic performance may also be overrated. For example, after the signing of the Montreal Protocol in 1987, many firms discovered they could successfully develop and patent new non-ozone-depleting substances as a new source of profits (Brack, 1995).

However, the involvement of business in environmental policy has also been criticized for a variety of reasons. First, the evolution of environmental controls based on self regulation, or the adoption of codes of conduct and voluntary measures, has been criticized as 'greenwash' by companies wanting to avoid command-and-control legislation by claiming to have taken evasive action (Friends of the Earth, 1995). One example of this is the so-called 'Monsanto Pledge'. In 1986 this US-based chemicals multinational pre-empted legislation limiting hazardous chemicals in the USA by announcing that it would reduce its airborne emission of certain chemicals by 90 per cent by 1993. Monsanto was able to enforce this promise, and so successfully deflected attention from the fact that the company was still a large emitter of these and other pollutants (Cairncross, 1995:187). Other research has also indicated that self regulation is futile when the ethical behaviour of some companies can be undermined by unethical companies ('free riders') that choose not to undertake voluntary measures (see Eden, 1996; Forsyth, 1997).

Secondly, the use of value-added, or 'green marketing' policies has been criticized for being short term, and also reflecting populist images of the environment rather than less popular but arguably more important actions. For example, in the UK between 1990 and 1992, the average 'green premium' paid for environmental retail goods fell from 6.6 per cent to 4.5 per cent apparently because of declining popular interest in green consumerism coupled with the effects of economic recession (McKinsey and Co., in Cairncross, 1995:183). The campaigning group Survival International, for example, criticized the environmental UK retailer Body Shop for allegedly representing indigenous rainforest people in a degrading and stereotypical way (Corry, 1994).

Thirdly, business participation in the regulation process may lead to a weakening of regulation because of the possibility of officials being offered jobs in companies after their work in regulation has ended. This process is called regulatory capture (Gouldson and Murphy, 1998).

Problems such as these refer to the ability of private companies to implement publicly-defined environmental policy objectives. Yet in addition, there are also problems associated with the impacts of private-sector investment in international environmental policy. Most obviously, problems include abusing differences between environmental standards in host countries compared with those identified internationally. Companies may also benefit unethically from the existence of corrupt or dictatorial regimes to avoid social responsibility in investment. More generally, however, international investment may also address environmental concerns and perceptions of investing countries, rather than the concerns and needs of host countries.

As discussed in Chapter 1, many so-called 'global' environmental problems such as anthropogenically induced climate change may in fact be far from global in terms of their impacts or perceived importance. The physical impacts of climate change – such as rising sea levels; disturbed weather patterns; and changed disease vectors – vary greatly between countries, and are also controlled by the vulnerability of different social groups to these changes. Furthermore, the perceived importance of different aspects of environmental change also varies between different countries and social groups. Commonly, the valuation of forests and certain types of biodiversity for aesthetic reasons in the North is not always shared in the South, and to assume that the loss of these represent a 'global' problem may in fact symbolize the extension of Northern opinions about environment to other areas of the world (Buttel and Taylor, 1994; Yearley, 1996). In terms of climate change, the assumption that the key objective of policy is to reduce greenhouse gas concentrations by itself, rather than develop the ability of some countries to adapt to change, or to achieve sustainable development in the South, may result in the proposal of solutions that may be less acceptable to some parties than the original threat posed by climate change (Gupta, 1997; Wynne, 1994).

The next section discusses the debates concerning different policy options for climate change in more detail. However, it is clear that harnessing private investment for environmental policy brings a variety of potential advantages and disadvantages. On the positive side, business involvement in environmental policy can increase technological performance, economic efficiency and speed of action. On the negative side, business involvement may also result in the weakening of regulation and the emergence of policy objectives driven by markets in industrialized countries. Foreign investment for international environmental policy objectives may also replicate the environmental concerns of investing countries rather than address more locally identified environmental objectives. There is consequently a need to manage the integration of business investment with environmental policy by guiding mechanisms that can increase the positive and decrease negative impacts.

Table 2.3 shows a variety of options available to governments to incorporate private investment in environmental policy. This table makes it clear that public–private synergy in environmental policy does not simply imply self regulation or increasing market forces, but the creation of incentives or partnerships that allow investors to achieve publicly-defined objectives. In the table, category 3 in the bottom left corner shows the extreme command-and-control form of regulation, and category 7 in the top right corner shows the other extreme of free market allocation of resources. Much debate concerns the options in the centre of the table, or which combinations of market freedom and institutional control and incentives allow the most effective achievement of both public and private objectives.

Table 2.3: Environmental policy instruments available to business and government

	Government	<————>	Market forces
Indirect intervention	1) providing education and information to consumers	4) providing infrastructure, but no market intervention	7) free market allocation of resources
	2) market-based instruments, eg subsidies/ taxes/ pricing (eg Polluter Pays Principle)	5) agreements or partnerships between government and industry	8) niche or 'green' marketing (responding to consumer demands regarding environment)
Direct intervention	3) legislation	6) permissions quota	9) long-term investments in market

Source: adapted from Janssen et al (1995:75)

International investment and climate change policy

This section now considers the implications of the preceding two sections for debates about climate change policy. In particular, the section describes the various arguments presented for and against harnessing international investment for climate change mitigation, and the associated debates concerning JI and the CDM. The section closes by discussing how far international investment may be used for purposes that support local development as well as global environmental policy. The section provides a foundation for later chapters that discuss the problems of technology transfer and investment in more detail.

The Joint Implementation controversy

The most direct way in which foreign investment has been discussed in the context of climate change policy has been JI, and the related concepts of AIJ and the CDM. JI in general terms may be defined as the possibility of emission reductions in one country being funded from, or otherwise arranged jointly with another country. The usual rationale for this is that, since it does not matter in global biophysical terms where greenhouse gas emissions are reduced, it is better to invest in reduction where abatement is cheapest. This is often assumed to be in Eastern European or developing countries.

JI in the context of climate change was introduced at the Rio Earth Summit in 1992. Article 4.2(a) of the UNFCCC stated:

> ...developed country Parties and other Parties included in Annex I may implement... policies and measures jointly with other Parties and may assist other Parties in contributing to the objective of the Convention...

Clause 4.2 applies to industrialized countries, and at the time there was considerable dispute concerning whether the term 'other Parties' might include developing countries. In addition, the UNFCCC did not attempt to define JI. Many developing countries opposed the adoption of JI – or other mechanisms of 'flexibility' such as emissions trading – because they were seen to be ways for industrialized countries to avoid taking responsibility for climate change mitigation within their own countries (Chapter 1).

Partly in response to these concerns, a pilot phase for JI was introduced under the name of AIJ at the first Conference of the Parties (COP) in Berlin, 1995. This allowed countries to undertake climate change mitigation projects cooperatively for a trial period 1995–2000. However, such investment was not to be credited against any potential future greenhouse gas abatement targets (QELROs). As a result of this lack of crediting, many observers believed that AIJ did not offer sufficient incentives for participation by Annex I countries (those countries likely to receive QELROs).

JI was finally approved as a flexible mechanism under the Kyoto Protocol of 1997, but was restricted to the countries of Annex I. In effect this means that JI now generally refers to investment by the wealthier Annex I countries of North America and Western Europe in the countries of Eastern and Central Europe, although JI investment between OECD countries is possible and there has been some talk of European companies investing in less energy-efficient US companies. The CDM was created as a similar – yet subtly different – flexible mechanism for investment in non-Annex I (usually developing) countries. However, it is important to note the controversies associated with JI/AIJ before proceeding to analyse the potential value of the CDM.

The most common criticism of JI has been that it is a mechanism of flexibility, and as such apparently allows Annex I countries to avoid taking responsibility for costly greenhouse gas abatement within their own countries (Gupta, 1997; Jepma, 1995; Parikh, 1995). Yet in addition there are a number of other associated problems.

The so-called 'baselines problem' of JI refers to the difficulty of establishing accurate baselines for measuring the impacts of JI/AIJ projects on carbon sequestration. There are serious methodological difficulties in estimating the emissions savings resulting from a JI project and what would have occurred without it (Riemer *et al*, 1997). This is the so-called 'counterfactual' problem of measuring the reduction of emissions rather than simply the emissions themselves. At present, intergovernmental organizations (IGOs) like the World Bank use evaluation systems based on opportunity cost between, for example, a 'clean-energy' investment, and 'business as usual' scenarios such as investment in a coal-fired power generation plant. Estimation is also difficult in the case of carbon sequestration projects like reforestation because the next best alternative may also absorb carbon. Indeed, the existence of one AIJ forest conservation project in a country may not preclude investors also conducting logging elsewhere in the same country.

A second related problem is the so-called 'cherrypicking problem'. This problem is the tendency for Annex I investors to select only those projects or locations that are cheapest, thus leaving governments or international agencies to address other more expensive projects (Heller, 1998). The usual approach of implementing bodies such as the United States Initiative on Joint Implementation (USIJI) is to ensure that local governments have agreed to the acceptance of proposed projects. However, this does not always mean that projects are acceptable to people living in the region, as indicated by the next concern.

Finally, many JI/AIJ projects have focused on sinks, or the sequestration of carbon dioxide by reforestation rather than on the transfer of badly needed EST to rapidly industrializing countries. Reforestation, or forest conservation projects, are often adopted because they are relatively cheaper than complex investment in industry, and also because they allow investors to benefit from an additional sustainable forestry business. However, many scientific justifications adopted for forestry-based projects have been challenged as simplistic or even false (see Cullet and Kameri-Mbote, 1998), and as an imposition of Northern environmental values on developing countries. Such projects also do not add to industrial technology transfer as demanded by the South.

As a result of the controversies concerning JI, many observers have proposed that JI be restricted in the form and location of projects it undertakes. Box 2.2 shows 12 typical suggestions that aim to make JI more effective both in mitigating climate change and encouraging local development. However, the experience of JI to date has indicated such restrictions may be against current practices. The next subsection discusses the experience of AIJ during its early stages, and then the section describes the potential offered by the CDM.

> **Box 2.2: Twelve proposed principles to guide Joint Implementation**
>
> 1. Independent evaluation should precede decision to move to JI with crediting.
> 2. A comprehensive legal framework should be in place before crediting begins.
> 3. Cost savings from JI should produce net climate change benefits, enabling greater emission reductions as well as lowering costs.
> 4. Projects should produce significant local environmental and socio-economic co-benefits.
> 5. Annex I commitments should be met 'substantially' through domestic policies and measures.
> 6. The role of sequestration (eg forestry) projects should be limited.
> 7. JI must be 'flexible, dynamic and must promote full participation' in the Kyoto agreement.
> 8. Emission saving estimates should be 'conservative' in the face of uncertainty. Additionally, JI must:
> 9. take account of the timescale over which emission reductions/savings are expected to endure;
> 10. promote capacity-building in the host country;
> 11. promote the transfer of appropriate technology;
> 12. be conducted 'with a high degree of transparency and public participation'.
>
> *Source: Goldberg and Stilwell (1997)*

The experience of Activities Implemented Jointly

Despite the concerns described in the previous subsection, JI/AIJ was rapidly adopted by a variety of Annex I countries including the United States, Norway and the Netherlands after the 1992 Earth Summit. The USIJI was established in October 1993, and had approved 25 AIJ projects by July 1997. These amounted to a total investment of $450 million, at a low per unit abatement cost of between $1 to $7 per tC (tonnes of carbon). In the same time period, the Netherlands had approved eight projects, Norway four, Germany three, France two and Japan one (Bush and Harvey, 1997).

Experience of AIJ projects during their early stages indicates, however, that investment has been focused on a narrow range of projects and countries. By mid 1998, for example, Africa had attracted only one Dutch forest-protection project in Uganda; a Norwegian fuelwood project in Burkina Faso; and a French hydroelectric project in Zimbabwe. The USIJI, for its part, had yet to accept a project in Africa, instead it had focused on Latin America. In three rounds of project evaluations (February 1995, December 1995 and December 1996) the USIJI had approved 12 forestry or land-use based projects, and 13 featuring renewable energy. Nineteen projects were in Latin America (17 in Central America), five in Eastern Europe (three in Russia) and just one in Asia (Indonesia – a carbon sequestration project through reduced logging).

In addition to USIJI projects, the GEF had by June 1997 financed 14 AIJ projects. All of these have focused on renewable energy or sustainable industrialization projects. Seven of these projects are in Asia (including two in China), two each are in Eastern Europe and Latin America, and Africa and the Middle East have one project each (in Senegal and Jordan – both for renewable energy). Costa Rica, however, has developed a reputation for encouraging JI/AIJ projects and helped facilitate investment by establishing a government office in 1994 and a 'Carbon Fund of Costa Rica', which aims to build a portfolio of carbon-offset projects.

In 1997, the World Bank also launched a similar scheme to AIJ called the 'Global Carbon Initiative', in which countries or private companies could invest in climate change mitigation. The Initiative provides investors with an organization structure and a choice of projects available for investment from around the world, usually involving forestry. The World Bank hopes that such investment may be credited eventually for offsetting QELROs, or be used in the short term by investors for public relations. Indeed, the Japanese car manufacturer Nissan has expressed interest in conducting such investment and hence claiming it markets 'climate friendly cars'.

The Clean Development Mechanism

The CDM was created at the Kyoto Summit as a third mechanism of flexibility alongside emissions trading and JI. The purpose of the CDM is, in brief, to allow Annex I countries the ability to achieve part of their QELROs through investing in climate change mitigation projects in non-Annex I countries. In general terms, the CDM is similar to JI because it offers an incentive to international investment for climate change policy. It differs from JI in that its focus is essentially North–South (rather than strictly within Annex I); and because its defining text makes no mention of the word 'sinks'. Indeed, the suggestion is that the CDM could be used for projects of more direct interest to sustainable development in the South, possibly including technology transfer.

The establishment of the CDM was a surprise to many observers at the Kyoto Summit. The main impetus for the CDM came from the Brazilian proposal in May 1997 for a 'Clean Development Fund' (CDF). Under the CDF, Annex I countries failing to meet their national obligations within a specific budgeting period would have to compensate for their failure 'by other means, such that the net effect will constitute a positive contribution to the global mitigation of climate change'. The proposed CDF also specified that (Yamin, 1998):

- funds from the CDF would be distributed in non-Annex I countries, 90 per cent for climate change mitigation and 10 per cent for adaptation projects;
- contributions from Annex I countries that had failed to reach national obligations would be calculated on a fixed scale of $20 per ton of carbon per year, to be multiplied against the number of tons by which the Party had been decided to underperform;
- in addition, it was also suggested that the fund should be allocated on the basis of $10 per ton of carbon to implement non-regret measures by non-Annex I Parties

However, at the Kyoto Summit, the proposal for the CDF was transformed into the CDM during the last day of negotiations. The purpose of the CDF was virtually reversed from being a punitive device for Annex I countries who had failed to perform, to being a device by which Annex I countries can achieve their obligations. Under the G77 proposals at Kyoto, it was stated that:

> the CDF will receive contributions from those Annex I Parties found to be in non-compliance with its QELROs under the Protocol. The CDF will also be open for voluntary contributions from Annex I Parties.

In contrast, Article 12.2 of the Kyoto Protocol states:

the purpose of the CDM shall be to assist Parties not included in Annex I in achieving sustainable development... and to assist Parties included in Annex I in achieving compliance with their quantified limitation and reduction commitments...

The difference between the proposed CDF and the actual CDM suggests that there may still be lingering resentment from some Southern negotiators that what was originally proposed to be a punitive device has become a flexible mechanism. However, the Kyoto Protocol made the CDM irrevocable, and an important part of future climate change mitigation. The final subsection discusses the potential for using the CDM for advancing investment that may satisfy the needs of the South in addition to the achievement of QELROs by investors.

Enhancing international investment for local development

The aim of this book is to assess possibilities for using foreign investment as a means to accelerate the transfer of environmentally sound technologies to developing countries. As noted above, most JI/AIJ activity is currently related to sinks and forest-related projects. The aim of this subsection is to question how far the emphasis on sinks may contribute to local developmental aims in developing countries, and how investment in technology-related projects may be increased.

The use of sinks in climate change mitigation is controversial (see Adger *et al*, 1998). The main problems identified with sinks are that the impacts of land-use policies on carbon sequestration are difficult to monitor compared with what would have occurred in the absence of such actions. Such impacts can easily be overestimated. Another problem is that addressing climate change by increasing sequestration is arguably avoiding dealing with the main cause of increasing greenhouse gas emissions resulting from energy use and industrialization (Gupta, 1997).

Yet in addition, there are also a variety of other problems associated with reforestation and sinks projects that are commonly not discussed at the climate change negotiations. For example, forestry projects are often justified because they contribute to biodiversity protection as well as carbon sequestration. In fact, much research in biodiversity has indicated that simple reforestation may not increase biodiversity, and that this may require long-term and complex seeding strategies that may be beyond the ability or budgets of many sequestration projects (Cullet and Kameri-Mbote, 1998). In addition, many other claimed benefits of forestry projects – including watershed protection or the diminishment of soil erosion – have also been shown to be simplistic, and often avoid many recent advances in environment and development research on local agriculture and land clearance (see for example, Fairhead and M. Leach, 1998; Ives and Messerli, 1989; G. Leach and Mearns, 1988).

There are also political consequences of forestry projects that are often not discussed. For example, in Thailand during the year 1991–2 the then military government attempted to reclaim land that belonged to it in north-eastern Thailand by forcibly resettling many thousands of farmers in Buri Ram province and planting trees in their place. This project was justified partly on environmental grounds, and known within Thailand as the 'Green Isarn' (or Green north-eastern Thailand) project. Yet it is also clear that such reforestation projects had additional political purposes, and it ended in conflict when some 10,000 farmers marched on Bangkok in protest (Pasuk and Baker, 1995:390).

Such local-scale factors are rarely considered when JI/AIJ are discussed as ways to mitigate climate change. As discussed in section 1.1(a), there is much evidence to suggest that many solutions to 'global' environmental problems may actually introduce local problems of a more immediate and damaging nature (see Rayner and Malone, 1998; Wynne, 1994). The reasons why the learning from local studies of environment and development are not always communicated to the international policy arena and are still a matter of research.

This book does not contribute towards understanding why so many consultants and agencies avoid the evidence of much research on the real need and impacts of forestry projects. It does attempt, however, to overcome this problem by suggesting solutions to climate change that will, we hope, be acceptable to both North and South. One key problem in harnessing international investment for local development is the current perception that industrial technological projects are high cost and difficult to implement in comparison with forestry-related projects. The point of this discussion is not to suggest that sinks projects should be avoided in general. However, the aim is to seek environmental policy options that address local concerns as well as those identified by international negotiations.

The rest of this book aims to find such mutually acceptable suggestions by exploring the possibility for using international investment for technology transfer and renewable energy development. In particular, the book assesses the potential advantages and disadvantages of encouraging further investment by vertically integrated companies in developing countries.

Summary

This chapter has introduced the main theoretical themes of the book by looking in detail at the implications of harnessing foreign investment in climate change policy. In particular, the chapter assessed the nature and role of the firm in public policy; the problems of privatization and public–private synergy; and the opportunities and pitfalls of investment to offset QELROs. The key conclusions include the following.

- Private-sector investment offers positive benefits on environmental policy because it accelerates the provision of infrastructure, technology and international standards of manufacturing. The vertical integration of firms is often a crucial process in the ability of the private sector to operate efficiently and achieve public policy objectives. International investment in high-cost, high-technology industries may ultimately be more beneficial to local development than attempting to develop new indigenous firms in these markets.
- However, there are also potential negative impacts of privatization, including monopolization, loss of national competitiveness and a reduced importance of external regulators to ensure that investment satisfies public-sector objectives. There is consequently a need to ensure

that private-sector involvement is accompanied by strong governance that can offer incentives to or regulate the activity of industry.
- For environmental policy, private-sector involvement may also lead to a weakening of environmental regulation and environmental policy objectives led by the concerns of firms and their markets. In international terms, this may mean a replication of investing countries' agendas upon developing countries rather than the identification of locally relevant objectives.
- In climate change policy, JI/AIJ and the CDM offer ways for using foreign investment for climate change mitigation. However, all are controversial because they may undermine North–South accord, and possibly lead to a weakened form of greenhouse gas reduction.
- Yet the potential is there to harness JI and CDM investment for climate change mitigation by transferring badly-needed technology to developing countries rather than the automatic insistence on forestry-related projects that may be ecologically overrated and politically unpopular. The achievement of effective public–private synergy in climate change policy therefore depends on asking *'which public?'* Some investment schemes for climate change mitigation may in fact be driven by developed country concerns and markets rather than the priorities of people in countries receiving investment.
- Integrating foreign investment and climate change policy therefore is not simply about increasing opportunities for private investors, but about which combination of market and regulatory forces can allow firms the greatest freedom to achieve profits while carrying out public-sector goals.

The rest of this book addresses the particular impacts of international investment for renewable energy development in industrializing countries.

Chapter 3

Decentralized electrification and climate technology transfer

Introduction

This chapter advances the discussion of international investment and climate change mitigation by assessing some of the practical options for integrating the two. Decentralized electricity development is one clear way in which international investment may transfer EST, such as renewable energy technologies, to developing countries. Yet both decentralized electrification and renewable energy may take various forms and require careful management in order to satisfy local development objectives.

Technology transfer is also widely discussed as the mechanism to encourage the adoption and eventual local manufacture of new technology. However, the term is complicated and may be divided into vertical relocation of new technology by investment, and horizontal embedding of technology through education and careful financial management. In this way, technology transfer is broadly related to the discussion in Chapter 2 on vertical integration of business. Under new incentives for international investment for climate change mitigation under the CDM, such vertical integration and vertical transfer of technology are likely to become more common.

This chapter assesses the potential consequences of international investment on technology transfer and the local development of renewable energy technology. Promoting international investment may accelerate the availability of new climate technologies. Yet inward investment in new renewable energy technologies may threaten the competitive standing of indigenous industries, and lead to dependency on foreign-owned less appropriate technology. Successful technology transfer depends in part on the ability of firms to gain commercial and competitive standing in local technology markets.

The chapter is divided into three main sections. The first focuses on new trends in decentralized and renewable energy, particularly in relation to rural electrification in developing countries. The second section summarizes the debates concerning the term technology transfer, and assesses some lessons learned in achieving successful establishment of technology through investment. Finally, the chapter discusses the approach taken towards technology transfer in the UNFCCC, and looks at the financial and institutional measures that have succeeded in building renewable energy development.

Decentralized electrification and renewable energy

Decentralized electrification is the development of electricity generation and supply devices that do not require large-scale grid extension or the construction of large power stations or dams to supply the grid. The benefits of decentralized electrification include the ability to supply electricity relatively quickly without the need for long-term and costly infrastructure, or for protracted negotiations with SEBs. Furthermore, decentralized electricity may also be more accessible to poor people in remote rural regions than conventional electricity sources.

Currently, much decentralized electrification has been achieved through the use of mobile diesel generators. But is it also possible to use renewable energy technologies for this purpose. Decentralized electrification may mean a variety of applications including stand-alone devices or minigrid supply systems over an area of some few square kilometres, rather than the extension of large national grids. This section defines various renewable energy technologies, and then discusses the applications and problems of rural electrification. The aim of this section is to provide essential background to the book's more general discussion of business investment and climate change policy. For a more thorough review of the technological aspects of renewable energy, see World Energy Council (1994) or Grubb (1995).

Defining renewable energy technologies

Renewable energy may be defined as those energy sources that do not involve adverse environmental impacts through exhausting finite resources such as coal or oil. However, it is clearly misleading to consider all forms of renewable energy as equally beneficial in environmental terms. PV technology uses solar radiation to create electricity with little impact on the immediate environment (although the manufacturing process may have some impact). This compares with biomass generators, for example, which may emit methane or other greenhouse gases (GHGs), and may require inputs of local vegetation that has to be regrown in order to keep environmental impacts neutral.

There is a common distinction between large and small renewable energy technologies. Large hydropower (dams), geothermal plants and ocean thermal energy conversion (OTEC) may be considered to be large renewable energy technologies requiring long-term investment and are generally connected to national grid systems. Small renewable energy technologies such as PV, wind turbines, biogas generators and microhydro devices may be used in a stand-alone, or off-grid capacity, although larger versions or groups of these technologies may also feed grids. Appendix II provides a basic description of different renewable energy technologies and associated applications.

A further common distinction suggests that large renewable energy technologies have social and political impacts often considered to be against the principles of sustainable development. Large dams in particular have been associated with major local environmental impacts of deforestation and flooding of land, and political impacts of resettling large numbers of villages. Critics have often suggested that the aims of constructing large dam projects include attempting to control or exploit areas of land historically outside the state's control. Classic examples of popular protests at large hydro development projects include the Narmada dam in India and Nam Choan dam in Thailand (Rigg, 1995).

As a result of such problems, it is difficult to refer to renewable energy technologies within one category, and to identify them as equally beneficial to the environment. While renewable energies do contribute to global environmental priorities by reducing GHG emissions compared with fossil fuels, they do not always support other, local, aspects of environmental or social policy, and indeed may sometimes contradict these.

The role of different renewable technologies depends on relative costs and availabilities of alternative fuel sources, and these vary according to location and the development of energy markets. In the United States, for example, most renewable energies are in the midst of a major, long-term transition from being 'advanced technologies' with only a peripheral market role to becoming mainstream 'technologies of choice'. However, in some developing countries, such as India, the role of biomass generation is far more important and in some cases may represent a high level of expertise.

There is also the question of whether renewable energy technologies are best used in terms of off-grid decentralized electricity generation, or in connection with local or national grid systems. It is commonly assumed that many renewable energies are best used off-grid, or as stand-alone generating devices in remote areas. This belief, of course, is highly simplistic and there are many ways in which renewable energy may be integrated into grid systems via large wind or PV arrays (embedded generation) or by replacing fossil fuels in traditional generators with renewable biomass. In addition, minigrids may be constructed to transmit small voltages of electricity from renewable energy sources to households within a village, or to connect several villages in one district. The capacity for renewable energy technologies to be grid connected therefore often depends partly on advances in transmission and distribution technology for remote regions (Barnett *et al*, 1982; Grubb, 1995; Hurst and Barnett, 1990).

However, the integration of renewable energy into grid systems is subject to a variety of constraints. Technical problems include the difficulties of ensuring that renewable energy technologies such as wind turbines have sufficiently long lifetimes, or reassuring authorities that wind or PV energy can be effectively integrated despite the variable nature of the source. Economically, grid connection also depends partly on the existence of large grids in zones where renewable energy is generated and, at present, the operation and maintenance costs of renewables is generally higher than alternative sources. Finally, there are also a number of political objections. In the United States, for example, extension of wind power has been opposed because of the impact on local bird populations, and by the visual impact of wind rotors and transmission lines (World Energy Council, 1994:196-9). The pros and cons of using renewable energy in rural electrification are discussed further in the next section.

The potential growth in different technologies may therefore depend on the future investment in particular applications, and the technologies with the greatest flexibility may grow most rapidly. For high-technology sources such as PV, this may be achieved most effectively by the vertical integration of multinational energy companies specializing in PV production. Vertical integration would imply the production, marketing and distribution of technologies by the same company and its subsidiaries in end-user countries. For other technologies, such as biomass generation or low-technology solar thermal devices, manufacturing may already be well advanced in developing countries. As a result, it may be more effective for these technologies to be advanced through a process of horizontal integration, or the increase in numbers of companies adopting these technologies, rather than the specialized pursuit of low costs by one or more multinational.

Rural electrification and renewable energy

Renewable energy may be used in both urban and rural areas, but can be particularly effective for decentralized rural electrification. Commonly, the term rural electrification indicates the extension of national grid systems to rural zones and a continued dependency on fossil fuel generation. Decentralized rural electrification, however, involves the provision of small-scale generating equipment to villages that does not depend on large-scale grid extension. Many renewable energy technologies may also be used in a decentralized way, and so the opportunity exists to integrate local development policy with environmental objectives.

Like renewable energy, rural electrification is often spoken of in simple, generally positive terms without reference to the various complex options available, or the potential negative impacts of some of these options. Rural electrification offers to provide immense local development benefits in terms of health, education and economic development, and also reduce macroeconomic strains on the economy through processes such as rural–urban migration. However, the traditional grid-extension approach to rural electrification has been criticized for a variety of reasons. Electrification, as in Thailand, may reflect the desire of the state to increase control over remote regions or resources held by ethnic or political groups that are perceived to be a threat to the state (Hirsch and Warren, 1998). Furthermore, grid extension does not always make electricity available to all rural dwellers, but only those with sufficient income to pay for it (in the short term at least), or those who live in villages close to the electricity supply (see Ramani, et al, 1993). Box 3.1 lists some of the traditional factors that have dominated rural electrification, and some of the criticisms of these principles by the World Bank.

Rural electrification is often considered to be costly and uneconomic. However, statistics suggest these fears may be unjustified. The World Bank (1996) for example, has stated that there are some 2 billion people worldwide without grid-supplied electricity. Many of these are already paying large amounts on services that require electricity, and the current expenditure on kerosene, car battery charging or other forms of local electricity may be no more than what is required to finance the long-term purchase of PV solar home systems.

Conventional approaches to rural electrification have also distinguished between so-called 'productive' and 'non-productive' electricity uses according to whether each use can generate cash income to pay for technology. By this logic, for example, local industry or retailing may be considered 'productive', yet schooling and public lighting may be considered 'unproductive' because there are no local systems to collect money to pay for such communal uses. However, increasingly evidence suggests that providing such communal electricity may also lead to the establishment of many new 'productive' industries that

Box 3.1: Traditional approaches to rural electrification and criticisms by the World Bank

Conventional utility strategies for rural electrification have assumed:

- rural electrification is planned centrally alongside other aspects of rural infrastructure;
- the approach adopted is area coverage rural electrification (ACRE), involving comprehensive networking to provide electricity supplies to as many customers as possible within a designated area;
- electrification is undertaken on the basis of extending the existing grid and the easiest options for extending this grid;
- isolated generation of electricity is better than multiple sources of electricity supply;
- electricity intensification (or the increase of the load factor on transmission lines) is better than increasing decentralized power sources.

Villages for rural electrification have been selected according to:

- proximity to grid;
- socio-economic or political ranking of villages;
- population size and literacy levels;
- existence of ongoing rural or strategic development programmes;
- availability of local resources for decentralized power generation.

However, in a 1991 survey of 63 utilities in 27 countries, the World Bank (1991) concluded that some key problems of rural electrification schemes in developing countries included:

Government and utility interactions: utilities often lack autonomy, resulting in a lack of responsibility and clashes of objectives between government and utility. In particular, when governments set tariffs, this often led to utilities having insufficient cash flows for operations.

Internal institutional relations: utilities are often constrained by national economic policies for a select number of preferred power plants supplying urban and industrial areas. Utilities subsidize rural power plants for being good for social development, but without adequate accounting procedures for recovering these costs.

Economic and financial issues: the failure to cover costs often led to a cash crisis within utilities, which impacted on maintenance, training and overall performance.

Financial management: bills are often left unpaid, and electricity usage unmonitored. 'Meter landlords', or unofficial suppliers of power from public systems are also common.

Staffing and training: utilities are either overstaffed or understaffed, with poor salary and career structures leading to a lack of skilled personnel. Salaries were generally set by government civil service pay scales, which could be as much as 50 per cent below the equivalent private sector level.

Sources: Ramani (1997:30), but see Munasinghe (1990) and World Bank (1991)

may in the long term provide sufficient income to pay for communal services. Consequently, much conventional thinking on finance for decentralized electrification may avoid the market that potentially exists (see Gregory *et al*, 1997).

Figure 3.1 illustrates two approaches to rural electrification using investment in renewable energy. The most direct route for new investment is to supply technology to the highest earners in villages through market penetration. The alternative is to allow indigenous technological development through long-term horizontal integration. Vertical market penetration may provide the fastest supply of electricity generation ability to rural areas, but it might also mean a lack of attention to some low-income inhabitants, and also a path dependency on imported technology. Developing indigenous technologies will take longer, and will require expertise and finance, but may lead eventually to indigenous industries that are well integrated into local economies.

Increasingly, the choice between integrating centralized grid extension, imported technology and rural indigenous technology is leading to the combined use of all three options. The concept of the distributed utility (DU) refers to the generation and distribution of electricity via a number of sources rather than through one centralized utility (see Figure 3.2). The DU concept is a skilful integration of small generation and storage and demand side management (DSM) using small elements (of between 500 kW to 5 MW). Under this concept, the central generating station will still provide the majority of energy needs of all

Figure 3.1: Bottom-up versus top-down modes of renewable energy rural electrification

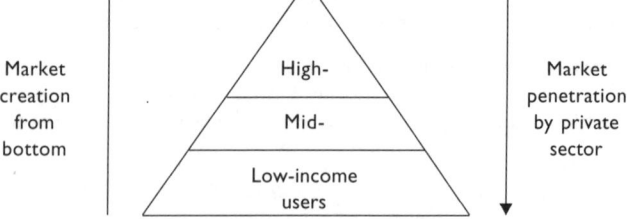

Sources: Ramani, et al (1993), Ramani (1997)

Decentralized electrification and climate technology transfer 51

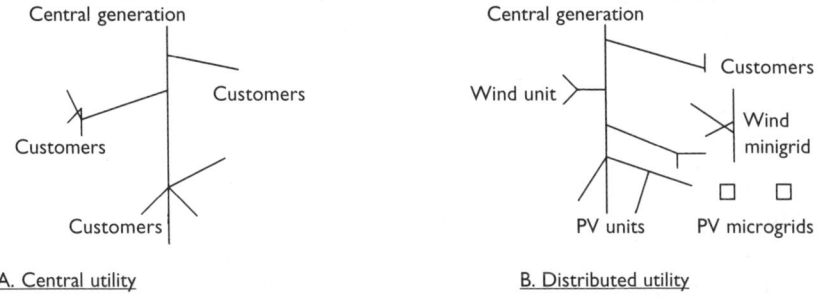

Figure 3.2: Central versus distributed utilities

Source: Iannucci and Eyer (1995:100)

customers, but the distributed elements will meet local demands in locations (such as remote rural areas) where there are infrequent peak loads, or where there are high upgrade costs.

Consequently, potential demand for foreign investment in renewable energy technologies depends in part on the ability to integrate these technologies into local grids and electricity demand patterns, and on the desire of the government to use decentralized power from international sources. Many of these requirements are political and may depend on the ability to break up central state bureaucracies in order to allow decentralized generation and planning.

Electricity investment and technology transfer

This section now addresses how international investment in new renewable energy technologies may lead to local renewable energy development in developing countries, and what this implies about the nature of technology transfer.

Defining technology transfer

Technology transfer is the diffusion and inculcation of new technical equipment, practices and development know-how from one region or company to another. However, the simplicity of this description belies the complexity of its meaning. Successful technology transfer requires attention to commercial, competitive and managerial aspects of business development, as well as ensuring the technical capability of new technologies in different locations.

The most common oversight in discussing technology transfer is an effective understanding of what is implied by both 'technology' and 'transfer'. Technology is often thought of as meaning sophisticated equipment, yet it can also include the skills, know-how and institutional support of relevant manufacturing or managerial resources. One classic definition is that of Schon (1967, in Leonard-Barton, 1990:45): 'any tool or technique, any product or process, any physical equipment or any method of doing or making, by which human capability is extended'. Consequently, full technology transfer can include training skills, management and general administration to ensure the successful adoption and commercial success of new technology. The word 'transfer' refers to the implementation of technical concepts into practice. Technology transfer therefore implies the process of upgrading existing production and technology development processes with newer and better forms of achieving production.

Many academic approaches to technology transfer make the distinction between vertical and horizontal technology transfer. Vertical transfer refers to the point-to-point relocation of new technologies via investment, often to a targeted group. Horizontal transfer describes the long-term process of embedding technology within local populations and economies. The distinction between vertical and horizontal technology transfer is important for the implications on technology ownership and location. Under vertical transfer, technology may still be owned by the investing company but it is being used in new locations. Under horizontal transfer, the ability to own and manufacture technology may fall into the hands of local users.

Technology transfer may therefore be compared with the process of business integration and the evolution of firms (see Chapter 2). Vertical integration in business theory refers to the enlargement of one firm, or the merger of two firms, to allow the new firm to undertake more than one stage of production at reduced costs (see Williamson, 1985). For

Decentralized electrification and climate technology transfer

technology transfer, vertical transfer or integration allows firms to expand the distribution of their products without having to risk the cost of sharing the ownership of technology. Horizontal integration in business theory refers to the establishment of contracts or joint ventures (JVs) with partners. In technology terms, it refers to cooperation with local companies in order to pass technological expertise onto them (see Box 3.2). Maximum profits should therefore in theory be associated with maximum vertical transfer (in order to prevent loss of intellectual property rights or costly JVs). However, many developing countries insist on the formation of JVs by international investors as a way to encourage horizontal technology transfer.

The division of technology transfer into vertical and horizontal, however, may also be simplistic as technology adoption also depends on the variety of technology offered, and the diversity of the host populations receiving it. Figure 3.3 shows four different modes of technology transfer. The main difference indicated here is the continuum between the number of individuals targeted as users for a particular application of technology, ranging from point-to-point – or a select number of potential users; to diffusion – or the widespread adoption of technology. In the case of renewable energy technologies some sophisticated PV systems may be best suited to point-to-point transfer because it is relatively more expensive and difficult to maintain than some traditional and locally manufactured technologies such as biomass or microhydro generators. Specific investment projects involving point-to-point technology transfer are most attractive to foreign investors because they imply the least contact with local organizations and end users. This diagram is eventually compared with renewable energy applications in Table 11.1, and with different regulatory regimes in South-east Asia, in Tables 5.1 and 10.1.

Traditionally, technology transfer is often discussed in terms of sharing expertise with local producers in order to enhance the developmental impacts of foreign investment. It is common, for example, for developing countries to require foreign investing companies to establish JVs with domestic companies in order to pass on the knowledge of outside investors to local producers (see Enos and Park, 1988; Ohkawa

> **Box 3.2: Vertical and horizontal integration in business theory and technology transfer**
>
> **Vertical integration**
>
> *Business application*: the extension of a firm's activities into different stages of production in order to reduce transaction costs. Integration may be either forward if it involves a manufacturer acquiring downstream distribution skills, or backwards if the extension is upstream. Vertical integration may also be achieved by a company merging with or taking over another firm at a different stage of production or by establishing a 100 per cent owned subsidiary outside its usual activities.
>
> *Technology application*: the point-to-point relocation of technology via investment without undertaking complex long-term education of techniques to local manufacturers. Vertical technology transfer is similar to vertical integration of business because it extends the activities of a firm without risking ownership of technology, or forming costly contractual relations with JV partners.
>
> **Horizontal integration**
>
> *Business applications*: (i) the extension of one firm's activities within its existing business niche by merging with or acquiring similar firms in that niche; (ii) the establishment of JVs or collaboration between different firms at either the same stage of production or at different stages. Taken literally, the application in (ii) is not usually considered to be 'integration' because it involves establishing costly contractual relationships between different firms rather than internalizing activities within one firm. However, in certain business circumstances (such as in some strictly controlled investment regimes in developing countries) such collaboration may be the only possibility for companies to extend activities. Furthermore, careful management of transaction costs under these conditions may also result in equivalent business economics to internalization.
>
> *Technology application*: the long-term embedding and establishment of new technology among local users and economies through training and financial management. Horizontal integration is often associated with training of local users in the potential applications and maintenance of technology, and in the commercial management of cost recovery. Eventually, horizontal integration may lead to the local manufacturing and ownership of the technology.
>
> *Sources: Stobaugh and Wells, 1984; Williamson, 1985; Leonard-Barton, 1990; Pitelis, 1993*

and Otsuka, 1994; Stobaugh and Wells, 1984). In particular, the ability to transfer technology depends on an effective legal network of patents, contracting and protection of intellectual property rights. Much research into technology transfer records the development of indigenous technology patents in new industries, and the numbers employed in technology production (eg Haksar, 1996; Lan and Young, 1996;

Figure 3.3: Four modes of technology transfer

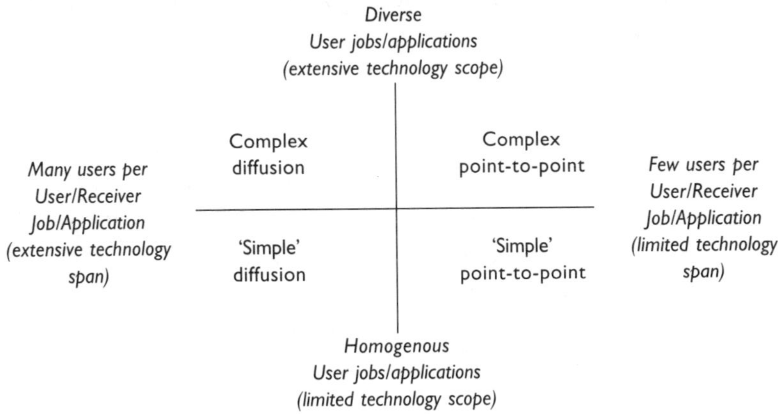

	Diverse User jobs/applications (extensive technology scope)		
Many users per User/Receiver Job/Application (extensive technology span)	Complex diffusion 'Simple' diffusion	Complex point-to-point 'Simple' point-to-point	Few users per User/Receiver Job/Application (limited technology span)
	Homogenous User jobs/applications (limited technology scope)		

Source: after Leonard-Barton (1990:47)

Markusen and Venables, 1996). Unfortunately, this kind of research can only be done retrospectively, and is therefore generally impossible for renewable energy exports where new markets are just emerging.

Complex JVs are a disincentive to investment in locations where companies are likely to incur high costs of market entry and then further costs of establishing local expertise (see examples of this in Chapter 7 concerning Vietnam). However, under conditions of vertical technology transfer investment may lead to the relocation of new technology without necessarily sharing this with local producers. The host country would then not gain competitive advantage within the production of that technology, but it would gain the associated benefits of investment such as taxes, employment and associated industries.

These debates have important implications for regional and national technology development strategies. As noted in Chapter 2, traditional approaches to technology policy have assumed a linear progression from research and development to innovation, to marketing and then dissemination. The aim of this approach has been to build national competitiveness in technology markets through protecting and investing in domestic industries. Under this model, technology is a commodity that gives companies or countries competitive standing, and therefore individual power in the market place may be developed by enhancing the development of this indigenous technology within a specific region or country (see Porter, 1990).

However, the linear model of technology development and transfer is increasingly challenged by those who argue that technology is dominated by international trade flows rather than national industrial policy (see Reich, 1991). The implication of this argument is that vertical integration of technology development may in fact be eventually good for overall technology and industrial development in host countries because it allows regions to become specialist producers of technology even if the technology is not locally owned. The first, linear, model argues that horizontal development within one region must at first precede competitive standing.

Successful technology transfer from foreign investment, then, does not simply depend on local practices adopted by investors or host communities, but also on national technology and competitiveness policy. Governments will repel potential investors if they are forced to share new technologies with local producers because this represents a threat to intellectual property rights and competitive standing. However, if governmenta are willing to encourage foreign investment without asking it to share technology, then the country may gain from the employment and development benefits of investment even if they do not gain access to the technology.

Improving technology transfer in practice

The theoretical discussion of technology transfer also needs to be accompanied by practical guidelines for successful transfers of EST from international investment to local users. Lessons for ensuring successful technology transfer projects in the future can be drawn from analysing the experiences of past investment projects. Martinot *et al*, (1997) identify three main channels for technology transfer. These are: technological perspectives (technical aspects of hardware and training suitable for different purposes); actor/agent perspectives (the intersection of activities by multinational banks and agencies, firms, non-governmental organizations (NGOs) and governments); and market/transaction perspectives (the market barriers and capacity for facilitating technology transfer as an economic activity). Success and failure of technology transfer can be attributed to the structures of one or more channel.

Box 3.3 provides some examples of poor technology transfer in renewable energy projects in developing countries. These provide incidences of failure on account of poor socio-cultural integration of technology with local users (India); inadequate identification of local needs for power (the Philippines); and weak competitive standing for high-technology projects (Brazil). As a result of these failings, the projects failed to gain acceptance by local populations in either technical or commercial terms. The examples also illustrate the differing need to consider technological suitability for target groups, as well as the competitive standing of indigenous industries when they are exposed to international investment.

Successful technology transfer at the local scale therefore depends on the correct alignment of local technological needs, and sustainable financial management (see MacDonald, 1992). These factors also have implications for official assistance with technology transfer. In general, problems with official assistance and cooperation have included:

- A tendency to subsidize technology rather than the institutions that aim to increase the adoption of technology at commercial rates.
- A lack of ownership of technology in developing countries has implied a dependency on the technology supplier for education and the supply of technology.
- A limited involvement of stakeholders in the technology identification, evaluation and training process.
- A dominance in external funding for choice of technologies, which has led to a limited choice of technologies.
- A tendency to overlook the need for a market 'pull' as well as a development or public-sector 'push'.
- A shortage of 'software', the skilled support staff linked to technology.

> **Box 3.3: Examples of poor technology integration**
>
> **The Fateh-Singh-Ka-Purwa biodigestors in India**
>
> This village in India was equipped with two biomass digesters by UNICEF in 1979 in order to provide biogas for local domestic cooking, electricity generation and agricultural machinery. The village was to provide operating costs. Unfortunately, the plant was decommissioned in 1985. The failings were technical and socio-cultural. In technical terms, the project failed because experts overestimated the supply of dung because they used national rather than locally gathered figures. In addition, villagers were reluctant to supply dung during certain agricultural periods of the year. In socio-cultural terms, planners failed to anticipate that the introduction of priced gas in the village for the first time immediately turned dung from a free good (collected by whoever needed it) into a traded commodity. As a result, arguments developed about ownership of dung that caused a shortage of supply to the digesters.
>
> **Biomass gasifiers in the Philippines**
>
> Gasifiers were widely introduced to the western Philippines between 1920 and 1949 as a way to power irrigation pumps. By 1986, 319 gasifier-engine-pump systems had been installed. Most used charcoal as a fuel and therefore needed a woodlot near plants to produce this. However, a survey at the time revealed that only a third of plants were operational, and that most were used only during droughts. The reason for this was that the pumps implied adopting a more intensive form of agriculture than that currently employed by most farmers, and that the pumps were seen to be less reliable than waiting for rain. In other words, the pumps were introduced because they were seen to be good business sense for increasing agricultural productivity in the long term. However, farmers did not currently desire this expansion, and adopted the pumps because they were offered free.
>
> **PV production in Brazil**
>
> During the 1980s the Heliodynamica company attempted to produce PV technology in one fully integrated and centralized factory. The aim was to sell PV to domestic and international markets. Almost immediately, the company encountered aggressive price competition from international investors who could increase technological standards at relatively low cost while Heliodynamica was fixed to its own technology. In addition, the company's ability to compete was affected by the use of most of its capital when the company was established, and it sought no further investments in production or partnerships with firms for either horizontal or vertical integration. The result of this competition was that Heliodynamica's product – although of high quality originally – quickly became dated in comparison with international imports, and the company became unprofitable.
>
> *Sources: Butera and Farinelli (1991); Gregory, pers. comm. (1998)*

- An inadequate institutional capacity to cope with technological cooperation (institutional capacity may be defined as the organizational infrastructure to educate local users, provide maintenance and long-term planning for technological development).

Decentralized electrification and climate technology transfer 59

- An inadequate appreciation of the risks of investors undertaking investment in technology.

These factors are discussed further in Chapter 3 and in relation to renewable energy in South-east Asia in Chapter 10.

Integrating renewable energy and technology transfer for climate change mitigation

This section integrates the preceding discussions of renewable energy and technology transfer by assessing what may be achieved in the context of climate technology transfer. The section first assesses the approach adopted to technology transfer within the UNFCCC and climate change negotiations, and then summarizes recent institutional and financial assistance for renewable energy development.

Technology transfer under the UNFCCC

As noted in Chapter 1, technology transfer is one of the most controversial aspects of the climate change negotiations. Developing countries have demanded climate friendly technology as a prerequisite for signing agreements, and as a way to avoid increasing greenhouse gas emissions during industrialization. However, developed countries have seen technology as privately owned, and hence largely outside the control of the state. Furthermore, many private investors have feared undertaking technology transfer because it is seen to be costly and a risk to intellectual property rights.

As a result of this impasse, there has been little progress in technology transfer, and debate has been dominated by traditional approaches suggesting that technology transfer is best achieved through donations at the governmental level, or by horizontal embedding led by ODA or public-sector bodies. This was indicated by one recent document: *The list of Chinese government needed technologies*, prepared by the Energy Research Institute of the State Planning Committee of the People's Republic of China (November 1996). This publication listed 15 specified environmental technologies including integrated gasification combined cycle; fuel cells; forest ecosystem management systems; and rice husk energy transfer instruments, yet there was no discussion of any compensation to producers. Technology transfer has become a symbol of the long-standing resentments between North and South about the responsibility for climate change, and has occasionally been discussed in emotive terms.

The emphasis on direct transfer of technology by governmental bodies was made clear in the wording of the UNFCCC and Agenda 21 agreements.

Article 4.1(c) of the UNFCCC (1992) stated that Parties should

> promote and cooperate in the development, application and diffusion, including transfer, of technologies, practices and processes that control, reduce or prevent anthropogenic emissions of greenhouse gases...

Article 4.5 stated that Annex II Parties

> shall take all practicable steps to promote, facilitate and finance, as appropriate, the transfer of, or access to, environmentally sound technologies and know-how to other Parties, particularly developing country Parties...

These statements indicated the great urgency of technology transfer, but did not make it clear how it could be achieved. Furthermore, statements in Agenda 21 suggested that technology transfer should be the main concern of the governments of developed countries, who should subsidize technology transfer in order to accelerate environmental protection. Chapter 34 of Agenda 21 (also 1992) suggested that the access to and transfer of EST should be promoted

> on favourable terms, including on concessional and preferential terms, as mutually agreed, taking into account the need to protect intellectual rights as well as the special needs of developing countries for the implementation of Agenda 21.

The words 'concessional and preferential' indicated that EST should be provided urgently and cheaply to developing countries through the

Decentralized electrification and climate technology transfer 61

actions of government and ODA, although the needs to reassure private investors about the protection of intellectual property rights was also acknowledged.

In order to achieve this technology transfer, Chapter 34 of Agenda 21 also urged the adoption of 'a collaborative network of research centres', and 'programs of cooperation and assistance'. These too suggested a dominant role of governments or intergovernmental organizations (IGOs). Since the UNFCCC in 1992, the first major initiative for centres was the Technology Assessment Panels (TAPs) set up through the Parties and Secretariat of the UNFCCC. TAPs sought to identify technology needs from developing countries in collaboration with investors. These panels, however, proved generally unsatisfactory because of disagreements about who should participate, and the extent to which technology could be transferred when it is privately owned.

A second major approach, the Climate Technology Initiative (CTI), emerged largely through the action of Japan operating through the International Energy Agency (IEA). The CTI is still evolving, but comprises a blend of voluntary actions by IEA member states; national technological plans; offering prizes for technological development; enhancing markets for emerging technologies; and collaboration between states on technology research and development. In addition, the CTI has also established regional seminars on climate change related technology, and has worked closely with the UNFCCC Subsidiary Body on Scientific and Technical Advice (SBSTA).

In 1997, the IEA announced that it would increase its work with the CTI and seek new collaboration with existing bodies such as the International Standards Organization (ISO). A new Global Remedy for the Environment and Energy Use – Technology Information Exchange (GREENTIE) initiative was also established, aimed at enhancing the use of climate change mitigating technology in ODA and private investment. Other initiatives include actions by the UN Commission for Sustainable Development (UNCSD) and UN Commission on Trade and Development (UNCTAD), which have sought to implement Agenda 21 through building cooperation and capacity in developing countries. Furthermore, international discussions now stress 'technology cooperation' rather than 'transfer' to indicate that technology transfer is not always meant to include loss of intellectual property.

However, the impacts of such official collaboration are as yet unclear. Furthermore, the problems in establishing the TAPs indicated that one of the most intractable barriers to accelerating technology transfer referred to the private ownership of much EST, and the problems in maintaining private investment in research and development when technology might be shared with potential competitors. Another factor was the relative stagnation of ODA in comparison with private investment flows. As a result of these trends and problems, the Kyoto Protocol of 1997 stressed the private involvement in technology transfer to a greater extent than Agenda 21 and the UNFCCC. Article 10(c) of the Kyoto Protocol stated (somewhat verbosely) that Parties should:

> cooperate in the promotion of effective modalities for the development, application and diffusion of, and take all practicable steps to promote, facilitate and finance, as appropriate, the transfer of, or access to, environmentally sound technologies, know-how, practices and processes pertinent to climate change, in particular to developing countries, including the formulation of policies and programs for the effective transfer of environmentally sound technologies that are publicly owned or in the public domain and the creation of an enabling environment for the private sector, to promote and enhance access to, and transfer of, environmentally sound technologies.

However, these approaches to technology transfer under the UNFCCC still imply a traditional linear model of technology development and dissemination. Instead, there is a need to redefine technology transfer under the UNFCCC in order to increase the relocation of some new technologies via vertical integration using foreign direct investment, and enhance the adoption of existing technologies from both North and South through horizontal integration. As discussed in Chapter 2, the CDM has this potential.

Some important clarifications are still needed before implementation of the CDM. In particular, it is unclear which guidelines will be used for establishing the added value of CDM projects on GHG emissions. Also, the United States and World Bank have made it clear that they

Decentralized electrification and climate technology transfer 63

would prefer CDM projects to be evaluated under the basis of existing AIJ frameworks; and also that existing AIJ projects be credited since their start date. Indeed, the United States has enquired if Annex I countries can also benefit from the CDM. If these proposals become law then there is little possibility of the CDM being used for technology transfer between developed and developing countries. There is consequently a need to demonstrate how foreign investment may be harnessed for technology transfer, in ways that do not threaten intellectual property.

Institutional and financial assistance for renewable energy development

Renewable energy has been the subject of a variety of official assistance schemes. There are two kinds of initiative: legal or regulatory changes to encourage private investment in renewable energy, and financial assistance to private investors from domestic, bilateral and multilateral sources.

The most discussed legal and regulatory reforms for renewable energy development have been the NFFO of the UK, and the PURPA of the United States (see section 1.1(b)). The NFFO (1980) was an accompanying legislation to the privatization of the electricity supply industry in the UK, and required electricity utilities to purchase a fixed proportion of electricity production from nuclear or renewable energy sources. The PURPA (1978) legislation in California broke the power generation monopoly and created markets for renewables by requiring utilities to buy power from qualifying renewable projects at the utilities' avoided costs.[1] These laws are generally held to have increased private investment in renewable energy and, as a result of this, also led to a decrease in the costs of renewable energy (see Chapter 1; Grubb *et al*, 1997).

---

[1] Avoided costs refers to the opportunity cost of purchasing energy from traditional sources. 'At avoided cost' implies that the new sources had the ability to supply utilities at or below this price level.

However, these new reforms also had unexpected effects. In the UK, the NFFO legislation resulted in damage to the British wind turbine industry as a result of strong competition from Danish importers. In the United States tax credits given to the solar energy industry were extended on a year-to-year basis during the 1980s, and then discontinued after 1987. This resulted initially in short-term investment only, and then a total lack of confidence in subsidization as a sustainable basis for building new industries (Kozloff, 1995a:12). Increasing competition may therefore have negative impacts on indigenous industries, but maintaining subsidies may create false markets that cannot last.

These experiences were in the UK and United States where electricity markets and regulation are relatively mature. However, lessons have been absorbed to varying degrees by official financial assistance programmes for renewable energy. Renewable energy first became an official priority for international aid at the 1981 UN Nairobi Conference on New and Renewable Sources of Energy, which called for action in research, planning, investment and dissemination of technologies. This identified a budget of US$5 billion (in 1982 dollars) for research and development until 1990. The Energy Sector Management Assistance Programme (ESMAP) of the World Bank was also established in the early 1980s as a donor-financed programme to develop energy sector strategies for poverty alleviation with the inclusion of renewable energy. However, resolve behind many new energy development programmes fell after energy prices decreased during the 1980s (Kozloff, 1995b:9).

Since then, the World Bank and the UN Development Programme (UNDP) have supported renewable energy development under a variety of projects. In 1988, for example the FINESSE (Financing Energy Services for Small Scale Energy Users) project was set up in order to incorporate new and alternative energy options into lending operations by the World Bank in the Asia region. However, actions have lacked coordination: some 25 agencies within the UN system alone have actively promoted renewable energy, with UNDP allocating US$50 million between 1990 and 1993.

Decentralized electrification and climate technology transfer 65

The World Bank initially concentrated funding on large-scale hydropower or geothermal projects. But in 1992 it established the Asia Alternative Energy Unit (ASTAE), with the mission of helping countries develop institutional capacity for building and adopting renewable energy technology and DSM measures. Activities have included setting up a variety of renewable energy projects in South and South-east Asia with a total value in excess of US$1 billion. The GEF, established by the World Bank, UNDP and UN Environment Programme (UNEP) in 1990, approved US$281 million for GHG reduction projects, DSM and energy supply efficiency. However, these have also included carbon sequestration projects that have not been directly related to encouraging the adoption of renewable energy technology.

In 1994, the World Bank also launched the Solar Initiative, which seeks to commercialize renewable energy use in developing countries. The initiative focuses on two programmes: an operational scheme to integrate new technologies commercially into World Bank and GEF projects; and a research and development programme that encourages a longer-term commitment to developing renewable energy at commercial levels. Despite its name, the Solar Initiative also provides support for wind and biomass applications.

However, much international assistance for renewable energy development has tended to focus on large-scale projects, and also has tended to overlook the practical needs for advancing technology transfer for smaller technologies. This is illustrated by the actions of the United States Agency for International Development (USAID). Between 1975 and 1988, USAID helped fund more than 200 renewable energy projects, but focused mostly on research and development rather than on an integrated process of developing then embedding technology in developing countries. USAID reviewed its practices in 1990, and suggested eight lessons for OECD donors (see Box 3.4).

Partly as a result of such lessons, the World Bank has developed new initiatives under its private-sector division, the International Finance Corporation (IFC), which can make direct loans to firms. These initiatives include the Renewable Energy and Energy Efficiency Fund (REEF).

> **Box 3.4: USAID lessons for institutional finance for renewable energy technology transfer**
>
> - Only commercially mature technologies should be used in projects that are not explicitly designed to promote technology development.
> - Only commercially competitive technologies – those that are affordable, reliable and easy to service – will succeed.
> - Local participation and market testing should be required as part of project design, implementation and evaluation.
> - The agency should address fuel subsidies and other policies that hamper the diffusion of renewables.
> - Technological applications should be tailored to fit a locale's social, economic, physical and institutional conditions.
> - 'After-sales' service must be adequate or renewable energy promotion will fail.
> - Local private-sector production, marketing, sales and service are needed to disseminate renewables and make a significant impact on a developing country's energy sector.
> - Better documentation of past experience could increase the rate of future success.
>
> Sources: USAID (1990), Kozloff (1995b:11)

REEF offers debt and equity finance to investors in renewable energy, and stipulates that 20 to 30 per cent of lending should be for off-grid technology, of which 20 per cent should be less than US$5 million in order to help small entrepreneurs. A second scheme is the PV Market Transformation Initiative, which seeks to award US$30 million to a variety of investments that advance PV usage. Finally, the Solar Development Corporation (SDC) will use funding from the IFC, the International Bank of Reconstruction and Development (IBRD) and GEF to operate as an independent holding company to market and support PV businesses worldwide (see Miller, 1998). These newer schemes place more emphasis on the power of the small entrepreneur in developing renewable energy, and on smaller technologies such as PV than historic funding that favoured large hydro and geothermal projects.

In addition to these developments at international level, there have also been successful projects by individual governments. In Argentina, for example, the Electricity Supply Program for the Rural Dispersed Population was undertaken in the 1990s in cooperation with participating provincial Regulatory Authorities (see Box 3.5). This

> **Box 3.5: Domestic finance for renewable energy in Argentina**
>
> The Electricity Supply Program for the Rural Dispersed Population gives priority to PV panels, small windmills, hydraulic microturbines, and diesel-driven generators. The total estimated investment of $314 million will be shared, with 45 per cent paid by users, 25 per cent from provincial subsidies and 30 per cent from national subsidies.
>
> The programme grants competitive concessions to one or more private enterprises in each province on the basis of lowest subsidy required per supplied user, technical qualifications and financial qualifications. The concession will run for 45 years divided into three periods of 15 years. At the end of each period the Regulatory Authority will call for a new bidding process, with the prevailing concessionaire having priority. Rates are negotiated between the concessionaire and Regulatory Authority for five-year periods. The concession shall be exclusive for users of up to 90 kilowatt-hours per month.
>
> As of late 1996, three provinces had at least begun the process of awarding concessions, with the remainder expected to do so by 2000. Two bids have been awarded. In each case there were four to five bidders, with a wide range in the bid values offered for combined rural and urban concessions. Concessionaires, who are established utilities elsewhere, are beginning with community applications in order to gain experience in their markets. The next stage will tender separate offers for the urban and rural markets, although the same bidder may bid for both.
>
> *Source: Fabris and Servant (1996), in Kozloff (1998)*

scheme demonstrates the ability to integrate renewable energy development with local electricity supply through careful use of funding. Individual countries may also impact on the relative costs of renewable energy development. In one study in the United States, the ownership and associated financial structure of wind energy projects was shown to have significant impacts on energy costs, with the most expensive option being Independent Power Producer (IPP) ownership with project financing. Public utility ownership with project financing was slightly (1 per cent) less expensive, while public utility ownership with internal financing was 12 per cent less expensive, and investor-owned utility ownership with corporate financing the least expensive option (29 per cent less expensive than IPP). The factors responsible for these differences are the cost of debt and equity capital, the fraction of debt in the capital structure and the amortization period (Wiser, 1997, in Kozloff, 1998).

For these reasons, financial and institutional assistance to renewable energy development has tended to address both private-sector investors and the governance structure, or utilities of the electricity supply industries. SEBs in developing countries in particular may be targeted to improve efficiency, training and financial management (see Box 3.1). An alternative could be to establish locally-based rural energy service companies (RESCOs) with a remit of identifying opportunities for DSM and DUs in rural areas, or in seeking greater integration of government and utility priorities. Concerning private investment, it may also be more fruitful to invest in the organizations that allow the horizontal embedding of new technologies, than in subsidizing the technologies themselves.

Summary

This chapter has discussed the theoretical implications of harnessing international investment for climate change policy by focusing on the practical options of decentralized electricity generation and technology transfer. It is argued that the employment of international investment for technology development will lead to greater vertical, or point-to-point technology transfer, rather than conventional horizontal transfer or long-term embedding usually associated with ODA. This new development will require attention to local and national institutions necessary to ensure such point-to-point transfer succeeds, and impacts on local indigenous technologies.

The key conclusions include:

- International investment may influence the implementation of decentralized rural electrification because it is an investment option that coincides well with international climate change policy and the availabilities of new technologies. Decentralized electrification also supports growing concern about the impact of large hydro projects; the cost of national grid infrastructure; and the bureaucratic inertia of many SEBs.
- The impacts of such inward investment on local technological development may be to reduce the competitiveness of local renewable

Decentralized electrification and climate technology transfer 69

energy technology industries, and potentially to encourage dependency on technologies that are not as appropriate as indigenous products.
- Current debates about technology transfer – particularly in the context of the UNFCCC – stress the need to develop long-term embedding of technology into local markets and social systems. However, this approach places high costs and commercial risks onto the private sector, and may avoid the benefits of so-called vertical technology transfer, in which new technology is relocated via inward investment without such local embedding. For renewable energy, benefits may include accelerating electrification and local development objectives, albeit at the risk of incurring conflicts with local technology manufacturers.
- There is consequently a need to understand more about how vertical technology transfer can benefit local development and may also be embedded locally in order to lead to long-term adoption of technology.
- A variety of official assistance schemes at international and national levels have existed to accelerate renewable energy development and rural electrification. Key requirements of such schemes seem to be incentives to encourage private-sector investment, and an acknowledgement that only those technologies that may be globally competitive should be supported. The PURPA and NFFO legislation of the United States and the UK offer potential models for building renewable energy investment elsewhere. Yet the evidence from countries undergoing privatization and liberalization of SEBs suggests that the encouragement of decentralized electrification also depends on the creation of decentralized governance or facilitating organizations that may operate independently of the SEBs.

The following chapters apply these conclusions to case studies in Southeast Asia in order to identify how international investment in renewable energy may be used for successful technology transfer.

Chapter 4

Electricity investment and privatization in South-east Asia

Introduction

This chapter concludes the introductory section of the book by reviewing current trends in electricity sector reform and investment in South-east Asia. South-east Asia is an appropriate region for analysis because, despite the financial crisis of 1997–8, it is rapidly industrializing and experiencing severe energy shortages that may be addressed by international investment. Large areas of many South-east Asian countries are without grid-supplied electricity and so decentralized rural electrification using renewable energy may be of great value.

The economies of East and South-east Asia have been the fastest growing of the twentieth century[1]. Between 1960 and 1995, Gross Domestic Product (GDP) in the so-called four 'Eastern Tigers' of South Korea, Taiwan, Hong Kong and Singapore grew at rates close to 9 per cent annually. More recently they have been joined in this growth by Thailand, Malaysia and China. In 1995, investment to Thailand, Singapore, the Philippines and Indonesia alone accounted for 5 per cent of all foreign direct investment. In 1995, investment in China amounted to 11 per cent of the global total.[2]

The impacts of such economic growth on GHG emissions are very large (see Ramani *et al*, 1992). South-east Asia is one of the world's major regions and home to some 500 million people, around 10 per cent of the world's population. The Asia Pacific region overall, including

---

[1] For the sake of this volume, 'South-east Asia' is taken to include those countries that are members of, or are eligible to join, the Association of South-east Asian Nations (ASEAN): Brunei, Cambodia, Indonesia, Laos, Malaysia, Myanmar (Burma), the Philippines, Singapore, Thailand and Vietnam. In statistical analysis, Cambodia and Laos have been excluded because of shortage of information.

[2] Unless otherwise stated, all statistics concerning GDP or energy/electricity generation and consumption come from IEA/OECD statistical reviews.

Electricity investment and privatization in South-east Asia 71

Japan and China accounts for about 25 per cent of global GHG emissions. In 2025 this is predicted to increase to 32 per cent (IPIECA/ UNEP, 1991, in Sharma, 1994:23).

The origins of GHG emissions in South-east Asia are complex. South-east Asia had one of the world's highest regional rates of deforestation during the 1980s at 1.2 per cent per year, compared with 1.3 per cent for South America, or 0.1 per cent for North and Central America. Furthermore, South-east Asia is dependent on irrigated rice production, leading to methane emissions from fields. Using figures for Asia Pacific as a whole, combustion of fossil fuels accounts for 36 per cent of regional GHG emissions; land-use changes (mainly deforestation) account for a further 33 per cent; and wet rice cultivation, energy mining and transportation add 26 per cent. CFCs account for 5 per cent (IEA, 1994, 1997a).

Since the financial crisis of 1997–8, Asian countries have been forced to reduce public expenditure on public infrastructure and environmental protection schemes. However, energy development remains one important way in which countries can build long-term economic development in order to avoid the impacts of recession. International investment for decentralized renewable energy development may therefore be a way for South-east Asian countries to increase electricity supply in rural areas, without needing to spend large amounts on grid extension.

The chapter is divided into three main sections. The first provides an overview of electricity investment in South-east Asia, and the growth of IPPs as major generators of energy. The section assesses the ability to integrate this investment with renewable energy development by looking firstly at available resources, and the current extent of decentralized electricity generation. The last section discusses the potential impacts of the South-east Asian crisis by outlining the nature of the crisis, and what impacts this may have on renewable energy and decentralized electrification. The discussions in these chapters provide a basis for the specific country case studies of Thailand, Vietnam, Indonesia and the Philippines in Part II of the book.

Electricity investment and private investment

Trends in electricity demand and supply

This subsection summarizes trends in electricity demand, supply and investment in South-east Asia. Figure 4.1 indicates the relationship between GDP and electricity consumption growth rates in Asia for the years 1986–95. These figures indicate that electricity demand is accelerating quickly in a generally proportional relationship with GDP growth.

For example, during 1986–95, electricity consumption in the OECD increased only by 3 per cent per year, but in Indonesia, Thailand and Malaysia the respective figures were 24, 22 and 19 per cent. China's consumption increased by some 12 per cent and Singapore's by 11 per cent. The average increase for South Asia during 1986–95 was 11 per cent. In Myanmar electricity consumption actually decreased during the same period by an average annual 6 per cent. These rapid growth rates indicate a new need for electricity generating capacity.

Figure 4.1: The relationship of GDP growth rate and electricity consumption growth rate in Asian countries 1986–1996

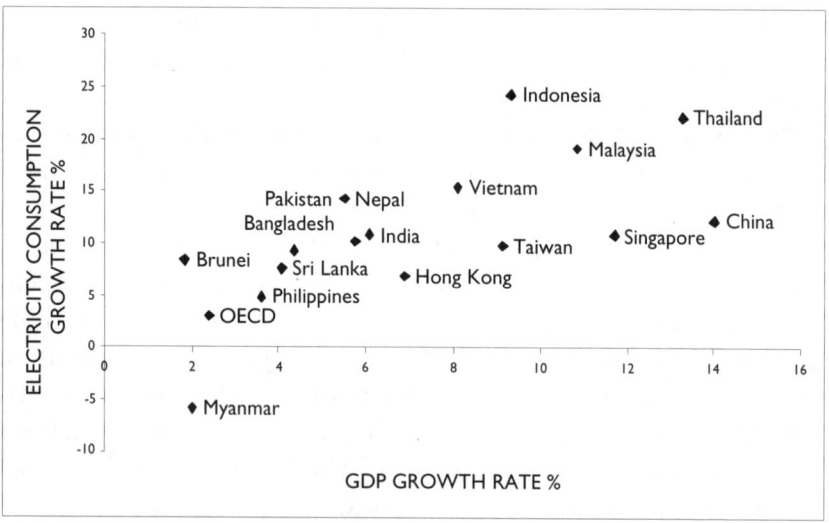

Source: *IEA/OECD energy statistic and balances of non-OECD countries, 1994–1995*

At present, the generation of electricity in Asia is heavily dependent on fossil fuels. Table 4.1 indicates the adoption of different fuels by various countries of Asia to generate electricity. The OECD, for example, depends on coal for nearly 40 per cent of electricity generation, yet in India this dependency is nearly 70 per cent. In Malaysia and the Philippines, however, the figure is less than 10 per cent, and each country instead depends much more on imported oil. Influences on the dependency on different fuel types include the existence of indigenous fossil fuel reserves, or the availability of large hydroelectric power (HEP) capacity in the form of large rivers and dams. In particular, Indonesia is the region's largest producer of coal, oil and natural gas.

Table 4.1: Generation of electricity by fuel source for selected Asian countries in 1995 (%)

	Coal	Oil products	Natural gas	Nuclear	Hydro	Other renewable	Combustible waste
OECD	37.4	7.9	12.8	24.3	15.8	0.5	1.4
India	69.4	2.9	5.9	1.7	20.1	–	–
China	73.4	6.1	0.2	1.3	18.9	–	–
Taiwan	33.7	26.2	4.2	28.7	7.2	–	–
Hong Kong	97.7	2.3	–	–	–	–	–
Indonesia	23.5	26.8	32.0	–	14.2	3.6	–
Malaysia	8.6	39.5	38.3	–	13.7	–	–
Philippines	6.8	62.8	–	–	10.9	19.6	–
Thailand	18.5	30.5	42.3	–	8.4	–	0.3

Source: *IEA/OECD energy statistics and balances of non-OECD countries, 1994–1995*

The influence of nuclear power upon future scenarios of electricity generation in Asia is unclear because of the obvious strategic implications such investment may bring. During the 1990s, nuclear power was concentrated within only a few Asian countries that possess the technology to produce it. Taiwan has been by far the highest producer of nuclear energy since the 1970s, with a contribution to total energy production of about 30 per cent. China is also growing in importance, with a contribution to power generation from nuclear fuel of less than 5 per cent. Chinese involvement in nuclear power is likely to increase, but at present there is little prospect of other Asian countries developing nuclear power capacity.

Investment in electricity generation by Asian countries is at a generally higher level than most OECD countries. Table 4.2 indicates the amount of investment in Asian power sectors as a proportion of GDP. Power's share of total investment in 1994 in the Philippines, for example, was 12.8 per cent compared with just 2.2 per cent in Germany.

Table 4.2: Power sector investments in relation to GDP in selected Asian and OECD countries

	Expected 10-year power sector investments (US$bn)	Proportion of annual investments as % of GDP	1994 gross capital formation in power (US$bn)	Power's share of total investment in 1994 (%)
Indonesia	50	2.9	59	8.5
Philippines	21	3.0	16	12.8
Thailand	33	2.3	56	5.9
South Korea	75	2.0	137	5.5
Germany	76	0.4	348	2.2
Japan	397	0.9	1,312	3.0
United States	430	0.6	1,194	3.6

Source: IEA (1997:78)

However, current levels of national investment and international lending may be insufficient to pay for all planned construction. The IEA (1997:77) has estimated that total investment necessary for power development in Indonesia, the Philippines and Thailand will be between US$65 and US$100 billion. However, total funding from multilateral and private banks was just $15.5 billion between 1984 and 1994; that national investment between 1990 and 1992 alone was some $12 billion (see Tables 4.3 and 4.4). Private-sector investment may therefore have to make up this shortfall. Figure 4.2 suggests that private-sector investment for all Asia (including South, South-east and East Asia) may potentially amount to 40 per cent of new power development between 1997 and 2020.

Table 4.3: National investment in power for selected countries 1990–2004 (US$bn)

	1990	1991	1992	1995–2004
Indonesia	1.55	1.46	1.42	82.0
Malaysia	0.52	1.12	1.21	17.8
Philippines	0.38	0.46	0.69	19.0
Thailand	1.70	1.88	2.21	49.0
TOTAL	4.15	4.92	5.53	167.0

Source: World Bank, in Breeze (1997:35)

Table 4.4: Total foreign lending to South-east Asian power sectors 1984–94 (US$bn)

Lending source	Japan Exim and OECD	World Bank / IBRD	Bilateral agencies	Asian Develop't Bank	Commercial Banks	Total lending
Indonesia	1.13	2.55	1.06	1.65	0.10	6.49
Philippines	1.78	0.62	0.87	0.38	0.32	3.97
Thailand	2.43	0.50	0.83	0.73	0.00	4.49
Share (%)	36	25	19	18	3	100

Source: National Power Utilities, reported in IEA (1997:76)

The growth in IPPs

As a result of this rapid increase in electricity demand, most South-east Asian governments have sought to attract private investment to provide funding and technology for specific projects of electricity generation. IPPs in general terms are power producers that are independent of electricity-purchasing utilities. They may be publicly or privately owned, but usually refer to private companies. IPPs may also sell directly to end users if they own transmission and distribution systems. The situation is occasionally confused by the ability of some utilities to buy shares in IPPs, such as in the Malaysian Tenaga Nasional, and therefore reduce the overall independence of IPPs.

Figure 4.2: Total planned cumulative capacity additions to Asian energy supply 1997–2020, with proportion available to the private sector (GW).

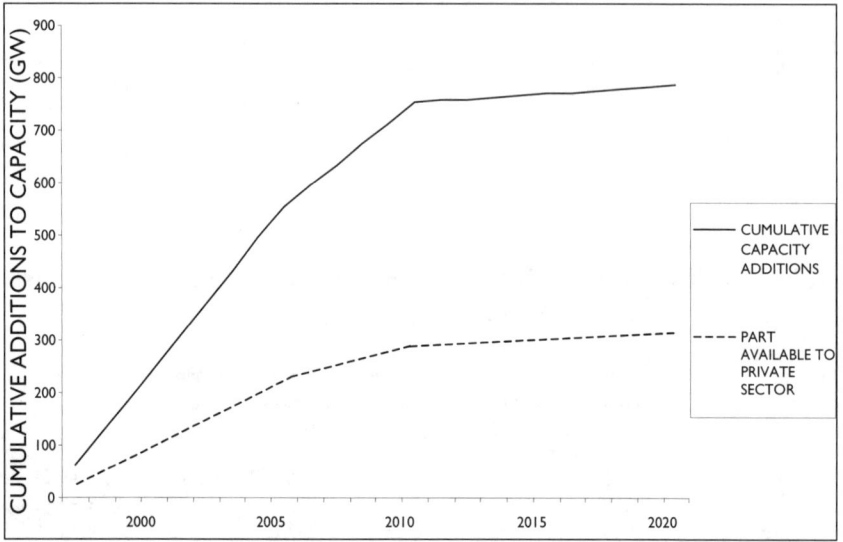

Sources: International Power Quarterly, and various trade journals, in Ripple and Takahoshi (1997:49)

As discussed in Chapter 1, the introduction of privatization does not necessarily imply an increase in competition (see Ripple and Takahoshi, 1997; USAID, 1998). Most IPPs in Asia have tended also to be large companies like Enron, PowerGen and Unocal, and have supplied grid-connected power from large power stations using fossil fuel energy sources. Privatization is therefore more likely to introduce a form of oligopoly to electricity supply rather than a free market for a variety of large and small companies. Furthermore, competition may not exist if the negotiations for supplying electricity are bilateral, or conducted under the provisions of an existing Power Purchasing Agreement (PPA). In general, the only competition that exists in the establishment of IPPs is in the initial bidding for the contract. After the contract is awarded, and the energy price agreed through a PPA, then there is no further competition that affects IPPs.

Electricity investment and privatization in South-east Asia 77

IPPs in the United States and Europe include both transnational companies and small and medium enterprises. Some IPPs also produce renewable energy. In Asia, virtually all electricity utilities are government owned (except for Japan and Hong Kong), and consequently there is less need for regulation. Furthermore, the main motivation for privatization has been the need to raise capital to construct the necessary infrastructure for projected electricity demand. To date, this drive to privatization has not been associated with an increase in competition. Almost all IPPs in Asia have been large or multinational companies, and have operated large-scale grid-connected power stations for base load supply, rather than locally-based projects including renewable energy.

Tables 4.5 and 4.6 indicate some of the IPPs already in existence in Asia at the beginning of 1997 and planned for the future. These figures show that countries usually characteristic for their cautious attitude to foreign investors such as Indonesia and Vietnam are attracting IPP investment. However, figures for IPPs under plan or solicitation have to be treated with some prudence following the likely reduction in some grand-scale energy investment schemes after the financial crisis, and the optimism of some governments in advertising some large-scale schemes even though they may not expect to receive funding.

Table 4.5: IPPs in operation at the start of 1997 in selected Asian countries (capacity, MW)

	No. of plants	Installed capacity (MW)	Average capacity (MW)	Smallest plant (MW)	Largest plant (MW)	Oldest plant (year)	Newest plant (year)
China	16	4,919	307	10	1,980	1987	1997
Malaysia	6	3,499	583	404	808	1994	1996
Philippines	18	2,743	152	7	735	1967	1996
Pakistan	1	1,292	1,292	n/a	1,292	n/a	1997
Indonesia	3	480	160	20	266	n/a	1996
Thailand	1	300	300	n/a	300	n/a	1996
Taiwan	2	69	35	33	36	n/a	1996
TOTAL	47	13,302	283	7	1,980	1967	1997

Source: Ripple and Takahoshi (1997:33)

Table 4.6: Private-sector power projects under construction, development or solicitation as of 1997 for selected Asian countries

	Projects under construction		Projects under development		Projects currently solicited	
	No. of projects	Capacity (MW)	No. of projects	Capacity (MW)	No. of projects	Capacity (MW)
Indonesia	10	10,150	25	9,942	n/a	n/a
China	16	5,866	74	80,869	23	36,943
India	8	5,023	82	56,751	28	96,967
Malaysia	4	4,589	8	4,645	6	740
Pakistan	17	3,611	14	5,319	7	48,672
Philippines	8	2,356	20	5,145	8	3,872
Thailand	6	1,515	10	3,217	n/a	n/a
Taiwan	1	1,350	5	4,660	4	13,780
Vietnam	n/a	n/a	14	5,402	7	11,165
Sri Lanka	n/a	n/a	7	8,705	6	9,640
Japan	n/a	n/a	2	200	6	2,720
TOTAL	70	34,460	261	184,855	95	224,499

Source: Ripple and Takahoshi (1997:53)

Opportunities for renewable energy development

Renewable energy resources in South-east Asia

Much attention to renewable energy development in South-east Asia has been dominated by large hydro development, and in particular to the untapped resources of the Mekong River (see Table 4.7). As discussed in Chapter 3, large dam projects are controversial for their local environmental impact and because of the large amount of resettlement needed. Many government projections for energy development include reference to large hydro projects (eg World Bank et al, 1996). However, the future of such projects is uncertain as a result of their cost and the likelihood of political protests over their construction (see Hirsch and Warren, 1998).

Decentralized renewable energy development, using smaller technologies such as PV, passive solar, wind turbines, microhydro and biomass generators may therefore be politically more acceptable, and also of greater access to poorer villages in rural areas. Figure 4.3 shows the

Electricity investment and privatization in South-east Asia

Table 4.7: Estimated and exploitable reserves of hydropower in South-east Asia and installed capacity, 1995 (in GWh and MW)

	Theoretical reserves (GWh pa)	Economically exploitable reserves (GWh pa)	Installed capacity (MW)
Cambodia	–	(8,400 MW)	1
Indonesia	400,063	18,172	2,169
Laos	–	(20,000 MW)	201
Malaysia	230,000	59,229	1,441
Myanmar	366,000	–	384
Philippines	46,759	18,814	1,959
Singapore	–	–	–
Thailand	555,500	9,088	2,459
Vietnam	>649	6,498	1,864

Sources: World Energy Council 1995 Survey of Energy Resources; Subregional Energy Sector Study for the Greater Mekong Subregion, Asian Development Bank

Figure 4.3: Proportion of total energy consumption coming from non-fossil fuel sources in South-east Asia 1972–95 (%)

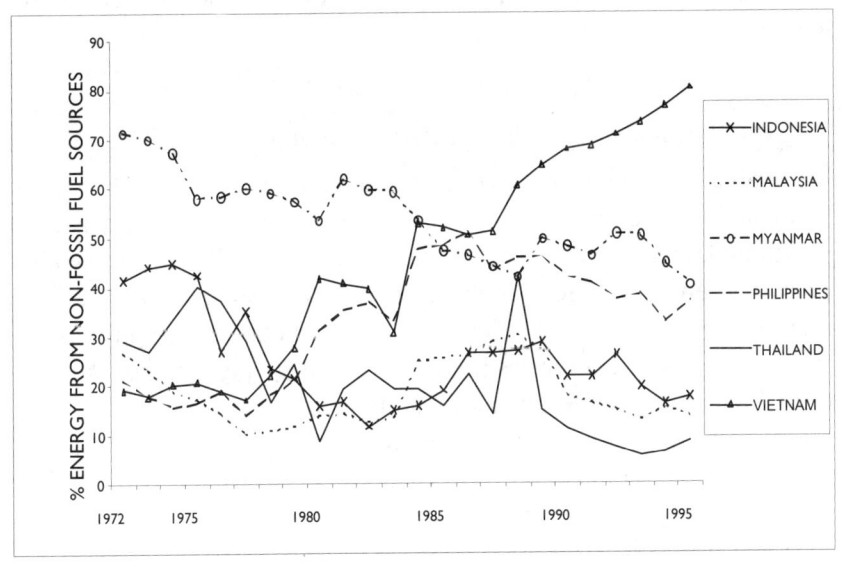

Source: IEA/OECD energy statistics and balances of non-OECD countries, 1994-5

proportion of total energy consumption coming from non-fossil fuel sources in selected countries of South-east Asia. These figures include energy from large hydro schemes, but refer to all energy uses in addition to electricity generation. The high dependency of countries such as Vietnam and Myanmar is an indication of the poor development to date of generation capacity powered by fossil fuels, and the existence of large hydro projects. In Thailand and Malaysia, for example, the low dependency on non-fossil fuels indicates a well-developed fossil fuel electricity generation sector.

There are few reliable long-term statistics about the availability of other forms of renewable energy resources, however these are widespread and offer great potential for electricity generation (see Ramani et al, 1993). Many developers of new renewable energy technology have complained that there are few reliable maps of wind speed and solar radiation available for South-east Asia.[3] This may also be true for biomass sources such as animal dung, crop residues and municipal waste, where inventories are difficult and alter rapidly as a result of wide-scale changes in urban and rural land use. Decentralized electricity generation may accelerate the adoption of such fuels for electricity generation.

Opportunities for decentralized electrification

Renewable energy development in South-east Asia has therefore been dominated to date by the construction of large hydro schemes, but the growth in decentralized electricity may increase the utilization of smaller renewable energy sources such as biomass and solar insolation. The potential for decentralized electrification in South-east Asia is huge. Large areas of most countries are without grid supply, yet the availability of finance for grid extension is diminishing.

Table 4.8 shows the proportion of populations in selected countries of Asia without connection to grid-supplied electricity in 1990. The

---

[3] Comments made at the Conference on Renewable Energy in Asia Pacific, Jakarta, October 1997.

table shows that for some countries such as Nepal and Myanmar, less than 10 per cent of the population are grid connected, yet for other countries such as Thailand and China, this figure is in excess of 65 per cent. As discussed in Chapter 3, the reasons for grid extension have not always included simple social development, but have included topics of national security and control. Also, the existence of grids in villages does not mean that all poor village members have access to electricity.

Table 4.8: Percentage of population with access to grid-supplied electricity in Asia, 1990

Country	% of total population (or villages where stated)	% of rural population (or villages where stated)
Bangladesh	12	9.74
China	66	79.79
India	80% of villages	25.00
Indonesia	24	22.00
Malaysia	82	80.00
Myanmar	6	5.93% of villages
Nepal	9	2.00
Pakistan	37	30.33% of villages
Philippines	61	54.00
Sri Lanka	29	18.00
Thailand	71	64.80
Vietnam	n/a	15% of villages

Source: Asian Development Bank (ADB), in Ramani, 1997:15

Decentralized electricity generation may refer to localized stand-alone or minigrid systems using technology such as PV or wind turbines, or small fossil fuel generators using diesel or fuel oil (see Chapter 3). Table 4.9 indicates the installed capacity of decentralized power generation in Asia for 1990. The contribution of such generation to national generating capacity is generally low, amounting to virtually nothing in Bangladesh and Thailand, and just 9 per cent in China. These statistics indicate that there is much potential to increase the importance of decentralised electricity to total generation. However, the importance of even the small levels in Table 4.9 to those villages with decentralized generation capacity must not be underestimated.

Table 4.9: Installed capacity of decentralized power generation in Asia, 1990

Country	Installed capacity of decentralized systems (MW)	Share of total national installed capacity (%)
Bangladesh	0.01	0.001
China	13,302.3	8.80
India	60.0	0.09
Malaysia	88.0	1.50
Myanmar	8.2	0.75
Nepal	1.0	0.38
Pakistan	2.0	0.03
Philippines	249.0	3.88
Sri Lanka	6.0	0.46
Thailand	3.0	0.03

Source: Asian Development Bank (ADB), in Ramani (1997:22)

Table 4.10: Electricity transmission and distribution losses in Asia, 1990

Country	T&D losses (GWh)	T&D losses (%)	Total system losses (%)
Bangladesh	2,597	35.6	39.1
Cambodia	73	41.8	44.7
China	40,113	6.9	12.9
Hong Kong	1,391	5.2	11.8
India	53,049	22.1	28.1
Indonesia	5,543	16.4	20.4
South Korea	5,620	5.5	10.2
Laos	37	16.7	21.4
Malaysia	2,369	10.8	16.3
Myanmar	705	27.7	28.8
Nepal	219	29.5	29.9
Pakistan	8,114	21.9	24.0
Papua New Guinea	72	11.9	11.9
Philippines	3,763	15.6	19.2
Singapore	530	3.6	9.1
Sri Lanka	–	–	17.2
Taiwan	4,996	6.0	9.7
Thailand	4,023	9.8	14.6
Vietnam	–	22.1	26.8

Source: Asian Development Bank (ADB), in Ramani (1997:6)

Electricity investment and privatization in South-east Asia   83

Finally, another incentive behind effective decentralized electrification is the need to invest in large-scale transmission and distribution infrastructure. Many electricity systems in Asia experience losses in power as a result of poor maintenance, outdated equipment or theft from unauthorized users. Decentralized electricity provides an opportunity to reduce expenditure on either the construction or upkeep of transmission infrastructure, and also reduces the losses of generated electricity experienced in centralized systems. Table 4.10 provides a list of different measured rates of loss for Asian countries in 1990. In some outdated systems, such as Cambodia and Bangladesh, total system losses can be as high as 45 per cent. In more modern systems, such as Singapore and Taiwan, losses are below 10 per cent. Research on privatization of electricity supply industries in developing countries, however, has indicated that private investment has tended not to supply transmission and distribution needs, and instead has focused on generation (see Berg, 1997; Bruggink, 1997; also see discussion in Chapter 12).

The Asian financial crisis

The nature of the crisis

In 1997–8, the major economies of South-east Asia – particularly Indonesia, Malaysia and Thailand – experienced major financial crises that led to a devaluation of currencies and overall loss of investment confidence. The aim of this last section is to discuss the importance of this crisis for international investment in South-east Asia, and to suggest possible impacts it may have on renewable energy and decentralized electricity development.

The origin of the financial crisis in Asia is complex. Indeed, one immediate comment might be to ask whether there was simply 'one' crisis, or several occuring in different countries, and the different times. One major reason underlying the crisis is poor financial management in countries such as Thailand where the government had allowed the financial services sector to expand without adequately controlling the growth of credit for high-value property developments or land speculation. In addition, the currencies of Thailand and Indonesia were kept pegged to the US$ at artificially high rates which failed to acknowledge market forces (see Dixon, 1998).

The result of these factors led to a startling fall in South-east Asian currencies in 1997–1998, a series of loans from the IMF, and the drastic reorganization of financial sectors in Indonesia and Thailand. In July 1997, the Thai Baht was freed from its traditional link with the US$ and immediately fell 23 per cent on international exchanges. Following this development, the Philippine Peso fell some 10 per cent, and Philippine stocks by 22 per cent; and Malaysian and Indonesian currencies suffered similar setbacks. The crisis in South-east Asia was then joined by a string of corporate collapses in South Korea.

Much initial analysis of the financial decline tended to use hyperbole and an overstatement of what the changes imply for South-east Asia. Instead, some analysts have urged caution about the long-term impacts of the changes, and suggested that South-east Asia might recover economic fortunes gradually in the manner of Mexico after similar economic crises in 1982 and 1994. On these occasions, serious balance of payments deficits and currency crises led to a period of serious readjustment in the 1980s, and relatively short financial reorganisation in the 1990s. Arguably, these steps have already been taken in South-east Asia. Two probable outcomes are that economic growth may return to South-east Asia, but at slower rates than during the 1970s and 1980s, and that greater competition will emerge between South-east Asian countries as each attempts to reduce costs and gain specialization.

It is also clear that the financial crisis of South-east Asia has also been a crisis of political governance. For years the financial sectors of Thailand and Indonesia have been dominated by bureaucratic state-operated organizations which have not reformed at the same rate as the economy, and which have often sought to support the interests of the governing elite rather than the interests of sustainable economic development for all. The downfall of President Suharto of Indonesia in 1998 after 33 years in power is one indication of the political changes that have accompanied the financial crisis. However, a more gradual political change throughout public sectors in South-east Asia is required in order to ensure that rapid economic growth can be managed more equitably and competently.

Opportunities and implications for electricity investment

For the energy sectors of South-east Asian countries, the impacts of the crisis are as yet unclear. International private investment in the region has been temporarily set back. Government funds available for large infrastructure projects have been reduced. However, the need for new infrastructure and secure energy and power supply have never been greater. Indeed, the government of Indonesia at least has stated that an efficient and modern electricity supply industry may be one of the most important ways for affected countries to escape the problems of recession and attract foreign investment in manufacturing again (Earle, 1998).

Under these circumstances, foreign private investment may be the most effective way to increase funding for badly needed electricity infrastructure projects. Yet under the conditions of crisis, investors are likely to demand terms that minimise risks of investment. Furthermore, the trend to low risk projects might dictate that schemes financed are those that are the simplest and least liable to additional costs. Such schemes may be fossil fuel generating capacity rather than more forward looking, and less immediately profitable schemes such as decentralized rural electrification or renewable energy development.

The financial crisis may therefore increase the trend towards privatization. However, the impacts on liberalization, or 'unbundling' of electricity supply industries are less clear. As discussed in Chapter 1, the privatization of electricity markets in developing countries has usually been associated with the search for new funding for infrastructure rather than the overall reform and decentralization of decision making. Indeed, if the crisis leads to the decision to accelerate the implementation of existing plans for fossil fuel generating capacity, the impacts may be to slow down the processes of liberalization and decentralization rather than facilitate a more general political reform. However, if the need to reduce costs means that existing SEBs are seen to be costly and bureaucratic, the process of unbundling may be accelerated in order to reduce barriers to private-sector investment.

If privatization leads to the unbundling of SEBs, the impacts on environmental and development policy could be twofold. Firstly, if unbundling is undertaken simply to enforce existing fossil fuel generating plans, the potential for integrating international investment with climate change policy may be lost. The second option is to integrate the processes of unbundling and privatization with the creation of a new regulatory body or policy objectives that attempt to direct investment into new and positive directions. Indeed, research in both industrialized and developing countries on electricity sector reform has indicated that privatization alone is unlikely to produce environmental or social benefits such as renewable energy development (Grubb *et al*, 1997; Berg, 1997; USAID, 1998).

Yet even with such political reforms, it is also likely that the financial downturn may make foreign imports of renewable energy technology more expensive. Higher prices will depress demand, and may also force governments to adopt indigenous technology development policies rather than invite investment from existing foreign technologies (see Chapter 2). If the financial crisis has the effect of making new renewable energy technologies more expensive, then the most likely outcome is a slowing down in their adoption in developing countries.

Renewable energy development and decentralized electrification may therefore offer a variety of different attractions and disincentives to governments in South-east Asia. On one hand, they provide an opportunity to avoid costly grid extension and a reduction in macroeconomic costs such as rural–urban migration. Yet in the short term, accelerating investment in fossil fuel generating sources may be most attractive as a way to avoid recession. Government policies during the crisis to maximize investment in alternative energy development may be crucial in determining which course is taken. The following case study chapters provide more information about the different privatization and investment programmes undertaken by four countries of South-east Asia.

Electricity investment and privatization in South-east Asia

The studies indicate four different scenarios of renewable energy development and economic regulation that may provide valuable lessons for policy in other developing countries.

Part II
Case Studies

Chapter 5

Introduction to the case studies

Introduction

Part II presents four empirical case studies to illustrate new trends in international investment, and renewable energy development and technology transfer in South-east Asia. The aim of these studies is to identify the ways in which international investment may lead to the successful adoption of new renewable energy technologies, and which range of incentives and institutional support may be necessary to achieve success. The studies particularly assess the role of privatization and liberalization in the encouragement of international investment and development of renewable energy.

South-east Asia was chosen as the study region because it provides a variety of opportunities for assessing international investment and electricity development. Despite the financial crisis of 1997–8, the countries are still in general terms undergoing rapid industrialization as well as privatization and liberalization of energy markets. The region also offers the ability to compare different countries in terms of various regulatory and privatization practices, and for their physical suitability for decentralized electricity development using renewable energy technologies.

The information in Part II comes from a variety of secondary sources such as statistical reports, government publications and policy reviews, plus a number of primary interviews with investors, government officials and other facilitators of renewable energy investment in the region conducted in 1997–8. In order to maximize the value of this book the case studies are organized to reflect different market structures and regulatory regimes rather than simple descriptions of different policies in each country. Such descriptions of electricity and renewable energy policies in South-east Asia exist elsewhere (eg Green, 1997). This book

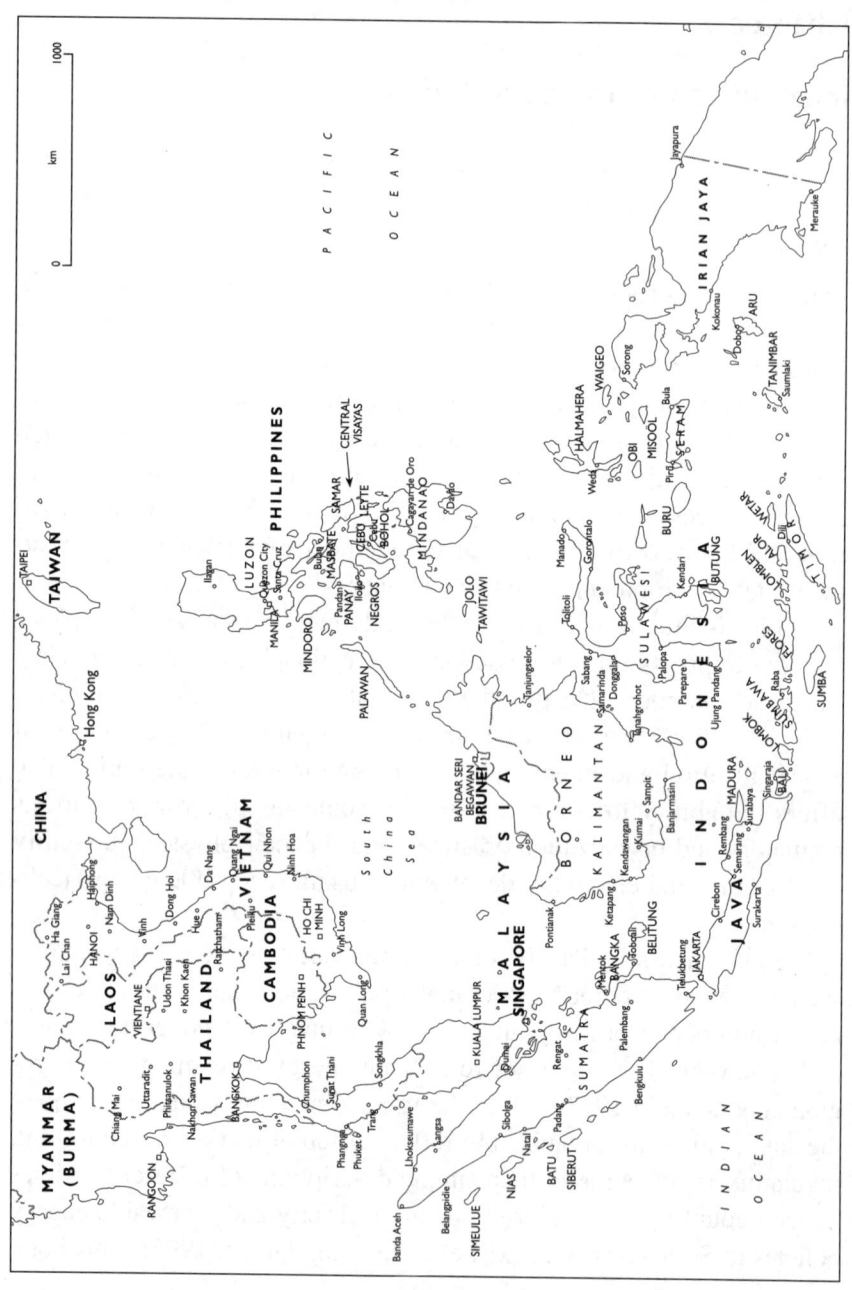

Introduction to the case studies

differs from these descriptions by attempting to analyse the various institutional barriers to international investment in renewable energy and technology transfer.

This chapter is a brief introduction to the case studies in Part II. There are two main sections in the chapter. The first section explains how the case studies were selected and suggests a basic classification of countries to differentiate market and physical incentives for renewable energy development. The second section describes the structure and aims of the case studies.

Selection of case studies

Four national case studies, of Thailand, Vietnam, Indonesia and the Philippines, are selected to demonstrate the impacts of international investment and privatization on renewable energy development and technology transfer. The aim of this section is to explain why these countries were selected and what the case studies seek to demonstrate.

Table 5.1: Classification of South-east Asia according to structures of business and regulation structures for renewable energy investment

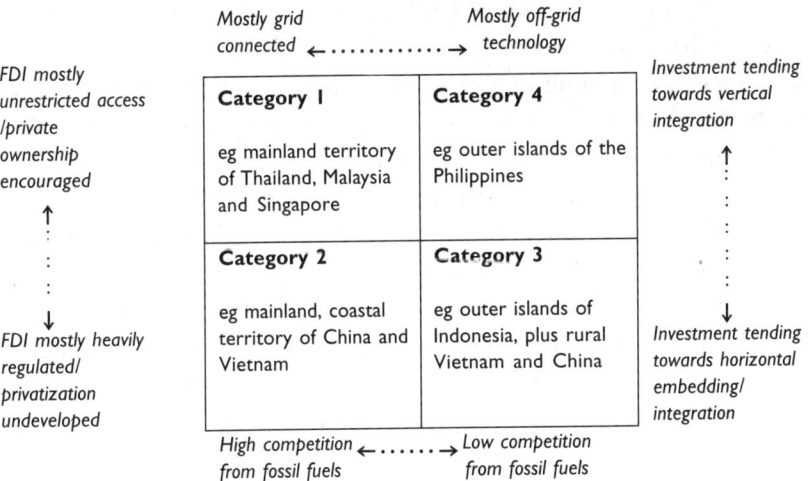

Source: the author

Table 5.1 shows a classification of locations according to the incentives available to renewable energy investment in terms of physical conditions and market incentives. The top and bottom axes indicates the physical suitability of locations for renewable energy, referring particularly to competition from fossil fuels and the coverage provided by national grids. The left and right axes indicate market incentives, including adoption of privatization and the openness of countries to point-to-point technology transfer from vertically integrated international companies.

The case studies in this book are organized into each of the four categories shown in Table 5.1. For example, Indonesia and the Philippines may be considered to be more suitable for off-grid renewable energy development because they have respectively some 15,000 and 7000 islands within their national territories. In terms of investment the Philippines are generally less restrictive for foreign companies than is Indonesia where foreign investors are more commonly required to forge joint ventures with local manufacturers. For this reason it is possible to refer to Indonesia and the Philippines in different categories.

However, it is important not to interpret this classification too rigidly as examples of each category potentially may be found in every country. Similarly, the grid-connected main islands of Java in Indonesia and Luzon in the Philippines may also be considered to be in separate categories to the outer islands.

It is also worth noting that renewable energy systems do not always have to be external to centralized grids, as suggested by Table 5.1. Much renewable energy development may be encouraged when it is supplied to national grids either from technology such as PV arrays or from factories with independent electricity generators that may sell surplus electricity to the national grid.

The aim of this classification, therefore, is to provide a guideline to those physical and market conditions that may provide the best incentives for harnessing international investment for local renewable energy development. The classification is not meant to refer to all locations within each country, or to suggest that countries may not move between categories following new policy initiatives.

Introduction to the case studies

Thailand, Vietnam, Indonesia and the Philippines were selected because they are large countries within South-east Asia that may represent the four categories, and because each is generally undergoing rapid industrialization. The less developed countries of South-east Asia, such as Laos, Cambodia and Myanmar are not included in the case studies because private-sector activity is still relatively undeveloped.

The case study chapters are therefore arranged around the following themes:

- *Chapter 6: Building renewable energy in grid-dominated areas: the experience of Thailand*
  This chapter assesses the difficulties of establishing renewable energy industries in locations where the national grids are already extensive and where these are already generally supplied by large centralized fossil fuel sources (category 1 of Table 5.1). Thailand was chosen because it is a mainland, and has a well-developed, grid-supplied electricity sector in which foreign investment is actively encouraged. In theory, this category is least likely to attract investment in renewable energy because of the dominance of existing fossil fuel supplies, and the reliance on market mechanisms rather than interventionist regulatory structures to allocate energy supply.

- *Chapter 7: Renewable energy investment under dominant state ownership: the case of Vietnam*
  This chapter assesses the pros and cons of renewable energy investment and technology transfer in locations where foreign investment is strictly controlled by governments, and where there is much domination by fossil fuel sourced grid networks (category 2). Vietnam was selected because it is still largely socialist in orientation, yet beginning to invite investment from outside sources.

- *Chapter 8: Renewable energy investment under strict bureaucracy: Indonesia*
  Indonesia is a good example of category 3, where renewable energy investment is encouraged by a large number of small islands,

yet where investment is still controlled by strict legislation and bureaucracy. In theory, this category may be expected to contain mechanisms to enforce horizontal integration or technology transfer, but also to contain barriers to unregulated foreign investment, or simple point-to-point vertical relocation.

- *Chapter 9: Off-grid renewable energy under active investment: the Philippines*
  This chapter concludes the different scenarios by focusing on off-grid systems under political systems where foreign and domestic investment is relatively open and encouraged (category 4). The Philippines are identified as an example here because of the high number of islands and remote non-electrified rural areas, but the role of bureaucracy and state intervention in the direction of energy supply is relatively less than in Indonesia, and energy markets are undergoing privatization. This chapter (category 4) should also provide a full comparison with Chapter 6 (category 1) by being – in theory – the scenario in which most foreign investment in renewable energy might be achieved in both vertical and horizontal terms.

However, it is also worth noting the various differences between each of these study locations in terms of economic development and current levels of wealth. Table 5.2 illustrates some of these economic and energy-related differences. In particular, it should be noted that the four case studies show great differences within South-east Asia as well as between Asia and the OECD in population size and per capita GNP as well as electricity generation capacity. The table confirms Vietnam as significantly poorer than the other countries mentioned, and also shows the difference between Thailand and Malaysia when compared with the Philippines and Indonesia.

Structure and aims of case studies

The case studies in Chapters 6 to 9 are arranged in order to reflect the different market and investment structures resulting from grid connection,

Introduction to the case studies

Table 5.2: Economic and social indicators for selected countries, including case studies, 1994

	Population (m)	GNP p.c. (US$)	GDP average annual growth 1985–94 (%)	Primary energy use (kg oil equiv. p.c.)[1]	Energy consumption average annual growth 1990–4 (%)	Electricity generation p.c. (kWh)[2]
Thailand	58.0	2,410	8.6	770	10.0	1,062
Vietnam	72.0	200	8.2	105	8.3	81
Indonesia	190.4	880	6.0	393	9.3	247
Philippines	67.0	950	1.7	364	8.3	400
China	1,190.9	530	7.8	647	4.0	711
Malaysia	19.7	3,480	5.6	1,711	11.2	1,713
Singapore	2.9	22,500	6.1	6,556	10.5	6,449
Australia	17.8	18,000	1.2	5,173	1.5	9,221
Germany	81.5	25,580	n/a	4,097	–1.5	6,693
Japan	125.0	34,630	3.2	3,825	2.3	7,211
Sweden	8.8	23,530	–0.1	5,603	0.2	16,913
United States	260.6	25,880	1.3	7,905	1.8	12,980

Notes: 1: Total energy use = domestic primary use before transformation to end-use fuels such as electricity
2: 1993 data for Asia, 1992 data for OECD

Source: World Bank and Asian Development Bank (in Symod, 1997:15)

competition from fossil fuels and degree of command-and-control regulation of foreign investment. These are organized in the format described in Table 5.1, and are generally at the national level with occasional reference to the sub-national case study.

The nature of analysis in each case study is an assessment of different institutional controls and opportunities for renewable energy investment and technology transfer. Assessing technology transfer following renewable energy investment in industrializing countries is extremely difficult to achieve. Much historical analysis of technology transfer has used econometric modelling of indices of technology transfer, or has focused on well-recorded statistical information providing histories of investment, patents and licences, employment or local participation in manufacturing activities (eg Coughlin, 1983; Khan, 1985; Lan and Young, 1996; Stobaugh and Wells, 1984; West, 1984; Yamashita, 1991).

Unfortunately for technology transfer and renewable energy investment, many of these trends lie in the future or have poor statistical records in developing countries. Other statistics used in economic modelling such as employment or local ownership of manufacturing may not also exist for many countries, or be unreliable. As a result, some well-established quantitative approaches to technology transfer research are generally not available for this study.

Furthermore, as discussed in Chapter 1, the aim of the book is also to indicate the institutional implications of international investment and environmental policy by analysing policy and market reforms. This requires a more qualitative and discursive approach than is usually adopted under the development and testing of economic models. As a result, the approach adopted in the case studies focuses on the nature of the market for renewable energy in different countries, and the legal and institutional framework adopted by three key actors of government, investors and intermediary organizations such as inter- and non-governmental organizations. It is noted, however, that there are several aspects of entrepreneurial activity and firm alliances that may be explained by social networks between different investors such as family interests, linguistic connections, and historic trading links (for example involving the overseas Chinese business community in Asia) (for example, in Europe see Charles and Howells, 1992). This book has not been able to research these networks, although longer-term, more detailed field research might reveal these (for example see Miller, 1998).

The case studies are therefore organized in order to indicate:

- The background to energy supply and renewable energy resources in each of the case studies. This section aims to indicate the potential for renewable energy development and the particular favouring of certain technologies over others. It will also indicate the competition for renewables from other energy sources.
- The structure and reform of the electricity sector. This section will describe the institutions administering electricity generation, transmission and distribution within case studies, and recent legislation for privatization and liberalization.

Introduction to the case studies

- Renewable energy development and technology transfer. This section will discuss the efforts that have been taken to encourage renewable energy investment within the case study regions (relating to vertical integration, or the introduction of new technologies via outside investment), and attempts to extend these new technologies to local communities and manufacturers via education, dissemination and institutional capacity building (horizontal integration).
- A final section will also discuss the policy implications of each case study, with attention to the pros and cons of building renewable energy in each case study location, and how particular policy measures have aided this process.

By adopting this structure it is hoped to indicate the institutional barriers and opportunities for investment in renewable energy and technology transfer under conditions of rapid industrialization. The implications of each category for achieving these objectives are discussed at the end of each case study, and then more generally in Part III.

Chapter 6

Building renewable energy in grid-dominated areas: the experience of Thailand

Introduction

This first case study looks in detail at the growth of renewable energy through investment and technology transfer in a country with a well-developed grid network and local availability of fossil fuels. This scenario corresponds with category 1 of Table 5.1, in which renewable energy development has to compete with existing cheaper sources, and where government policy favours price competition for energy development from fossil fuel sources. In theory, this first category may be the most difficult market structure for foreign investment to build a thriving and widespread renewable energy industry on.

Thailand has been selected as the focus of this case study. Unlike the Philippines or Indonesia, Thailand has few islands and is mainly composed of one large mainland. This has made it relatively easy to extend the national grid to every region, and in December 1994, 98.2 per cent of villages and 86 per cent of rural households in Thailand were connected to the grid. This has decreased immediate demand for decentralized rural electrification using renewable energy technology. Thailand also has local supplies of lignite and natural gas in the short term at least.

Thailand also fits the description of category 1 in Table 5.1 because it adopted a policy of energy privatization and liberalization during the 1990s that has increased the efficiency of electricity generation through exposure to market forces. Under these conditions electricity generation has favoured cheapest options, and therefore has made new renewable energy technology less competitive with fossil fuel sources. Yet despite these barriers to the adoption of renewable energy in Thailand, there have been initiatives to build renewables with some limited success,

Building renewable energy in grid-dominated areas

particularly in coordination with DSM. The nature of this success is worth observing in order to demonstrate how renewable energy may be built under these apparently difficult market conditions.

This chapter starts by describing energy trends in Thailand, and the nature and structure of the electricity supply industry. Measures for privatization and liberalization are then described before an analysis of the implications of the electricity restructuring on renewable energy investment and technology transfer. A final section then draws conclusions from this experience regarding the book's more general theme of seeking successful regulatory and institutional structures for integrating business investment with global environmental policy. The chapter's aim is to focus on the evolving relationship of market structure and renewable energy development, and therefore many general details of Thailand's energy trends and resources are not discussed here.

Energy supply and renewable energy resources in Thailand

Overview

Thailand is located centrally within continental South-east Asia sharing borders with Malaysia in the south, Myanmar in the west and Laos and Cambodia in the east. It has an area of more than 513,000 km^2. The country is commonly divided into four physiographic regions: the mountainous north, the central plains (including the capital Bangkok), the north-east, and the southern peninsula. These regions have great physical and cultural variety, ranging from deep river valleys and scattered mountain villages in the north; the predominantly poor and highly agricultural plateau of the north-east; and the forested and coastal landscape of the south. Most of the population is located in the central plains, which is also the most economically developed region.

Thailand's climate is also varied between these regions. In general, the country has a subtropical monsoonal climate with a hot season (March–May), rainy season (June–October) and cool season (November–March). Seasons may be late or early starting, with rainfall being unpredictable. Temperatures generally range from 24 °C to 30 °C. There are occasional typhoons in the Gulf of Thailand.

Thailand has been one of the fastest growing economies in Southeast Asia. Between 1985 and 1994, GDP rose at an average annual rate of 8.6 per cent. GDP per capita reached US$2,240 (or $1,822 in 1988 dollars) in 1994, with the consequent development of a large middle class (IEA, 1997:257). Agriculture accounts for about one-sixth of GDP, but still employs more than half of the workforce (Green, 1997:101). About one-quarter of Thailand's population of 60 million are classified as urban.

The rapid growth of Thailand's economy was largely a result of the Thai government aggressively encouraging foreign investment during the 1970s and 1980s, and by the country's abundant resources and proximity to other markets. However, the attraction of investment has arguably not been matched by a growth in stable government. Military governments were predominant until Thailand's first democratic elections in 1988. In 1991 a further military coup led a year later to a massacre of pro-democracy protestors in Bangkok. Thai governments have since been democratically elected, and passed reforms to liberalize the economy and increase other measures such as environmental protection. However, continued mismanagement of public borrowing and exchange rates have been blamed for the recession in Thailand since 1997, and this has reduced the middle classes' spending power, slowed foreign investment and led to tighter monetary standards. Large imbalances still exist within Thai society and between different regions. But the economic policies adopted since the recession make redressing these differences very difficult.

Energy trends

Thailand has a variety of natural energy resources, including natural gas and lignite. Table 6.1 indicates the extent of known and potential reserves of energy resources in the country. Natural gas production amounted to 17.1 million cubic metres per day in 1990, but is expected to peak before 2000. Crude oil and condensate production was about 16 million barrels in 1990, and is expected to remain broadly at this level. Some 8.9 million tonnes of lignite were produced in 1990, but this total is expected to rise to 36 million tonnes in 2001. In 1994, the total energy supply to Thailand was placed at 43 per cent natural gas, fuel oil and diesel at 26 per cent, coal 22 per cent and hydroelectric power at 6 per cent. Other sources of energy, such as small renewable power technologies and biomass were estimated at 3 per cent (Green, 1997:101–4).

Table 6.2 provides an estimate of energy supply trends in Thailand for 1995–2010. During the late 1970s, Thailand was dependent on imported oil for about half of its domestic demand for energy. Despite

Table 6.1: Coal, oil and gas reserves in Thailand, 1995

Resource	Unit	Proven reserves	Probable reserves	Possible reserves	Total reserves
Natural gas, total	bn cu ft	7,119.8	8,635.0	11,095.9	26,850.7
Onshore		624.0	129.3	229.1	982.4
Offshore		6,495.8	8,505.7	10,866.8	25,868.3
Coal, lignite	m tonnes				2,331.4
Crude oil, total	m brls	128.7	61.5	141.1	331.3
Onshore		75.1	49.2	43.3	167.6
Offshore		53.6	12.3	97.8	163.7
Condensate (offshore), total	m brls	166.5	190.7	109.2	466.4
Oil shale, total	m tonnes				18,500.0

Source: Thailand Department of Energy Development and Promotion, (adapted from Lefevre et al, 1997b:17–19)

Table 6.2: Predicted primary and final energy demand in Thailand 1995–2010, Mtoe (Million tonnes of oil equivalent)

	1995	2000	2005	2010
Total primary energy supply	61,144	89,262	112,071	137,021
Total final energy consumption of which:	34,120	48,935	64,823	83,133
Coal	2,004	2,686	3,294	4,805
Oil	24,542	31,513	41,982	54,158
Gas	931	1,996	2,590	3,105
Electricity	6,642	12,740	16,957	21,065

Source: IEA (1997:262)

increases in natural gas and lignite production within Thailand, dependency on imported energy is expected to reach some 60 per cent in 2001. From 1996 on, imported oil will account for roughly two-thirds of commercial energy supply. Demand for imported coal is also expected to increase, reaching some 6 per cent of power generation in 2001 (IEA, 1997:262; Green, 1997:105). The demand for power is therefore unlikely to be met through domestic production of resources (see also Table 6.3).

Renewable energy resources

Thailand has a large domestic potential for hydropower generation. An estimated 8,027 MW exist within Thailand, plus further even greater potential from the Mekong and Salween rivers on the Laotian and Myanmar borders upon which large internationally managed projects have been proposed. Presently, only 2,429 MW have been exploited within Thailand, with 516 MW being developed and 1,056 MW planned before 2006 (Green, 1997:104). Furthermore, the option exists to purchase hydroelectric power from neighbouring countries, notably Laos and southern China.

However, the construction of dams has been politically controversial in Thailand, and representative of wider political struggles within the country. In 1988, a proposal to build a dam at Nam Choan in Kanjanaburi province (western Thailand) was cancelled following popular protest at the destruction to forest and relocation of people the dam would have brought. Further protests have been associated with other projects such as a Pak Mool in the north-east of Thailand. The construction of large hydroelectric projects in remote areas of the country, often in locations where there has been historic insurgency, has been seen to be an attempt to control these areas by the Thai state. Furthermore, the construction of dams has become a symbol of environmental resistance amongst Thailand's growing environmental movement (Rigg, 1995; Hirsch and Warren, 1998). Therefore, the construction of large hydro projects is associated with political conflicts that make any future major commitment to hydropower unlikely.

Smaller renewable energy technologies may have more potential. Biomass continues to be of great importance. In 1988, 22.6 million tonnes of wood, 3.09 million tonnes of paddy husks and 7.2 million tonnes of bagasse contributed 33.29 per cent of Thailand's primary energy supply (Green, 1997:113). However, in 1988 the supply of wood for biomass energy was affected by the passing of a logging ban making it illegal for anyone in Thailand to log trees for any purpose without official permission. Nevertheless, the increasing commercialization of agriculture in Thailand creates a large supply of agricultural residue that may be used in industries such as rice milling as well as domestically within villages.

Mountain streams in the north of Thailand are possible sites for small hydropower generators, particularly as these mountains are also home to several small villages disconnected from main grid systems. Daily total solar radiation in Thailand has an annual average intensity of 17 MJ m^{-2} or 4.72 kWh m^{-2} day^{-1} (or equivalent to the heat released from burning 1 kg of dry wood) (Green, 1997:108). Wind speeds in Thailand are greatest along the southern coasts for either the Gulf of Thailand or Andaman Sea (Indian Ocean). In Phuket, observed maximum monthly wind speeds have varied between 7.9 m s^{-1} and 19.3 m s^{-1} (Green, 1997:118). There is little potential for geothermal energy as Thailand is not located on an active volcano zone. Nevertheless, there are some 90 hot springs with surface temperatures of 25–100 °C in the northern, western and southern parts of the country, with five of temperatures of more than 175 °C in the north. There is one small geothermal power station connected to a local grid in Fang, in Chiang Mai province in northern Thailand, with an installed capacity of 300 kW.

Structure and liberalization of the electricity supply industry

Electricity market and institutions

The generation of electricity in Thailand is characterized by the intensive use of fossil fuels to supply a large national grid system. Table 6.3 indicates the current and projected future installed electricity generating capacity in the country, with large increases marked for imported coal, oil and liquefied natural gas (LNG).

Electricity generation capacity has grown rapidly in Thailand. In 1970, the total installed capacity of the Electricity Generating Authority of Thailand (Egat) was just 1,258 MW. At the end of 1996, the total installed capacity was 16,141.5 MW. The government's Power Development Plan (PDP) (version 1997) aims to install a further 36,755 MW by 2011 of which new power plants would produce 13,640 MW (Lefevre et al, 1997b:57). Some 60 per cent of this additional capacity is planned to come from thermal sources, including oil, gas and coal (lignite) (see Table 6.4). Of the remaining 40 per cent, 18 per cent (6,441 MW) is planned from combined cycle; 7 per cent (2,470 MW) from hydro; 3 per cent (1,000 MW) from gas turbine; and 15 per cent (5,631 MW) would be purchased from Laos.

Table 6.3: Thailand: power plant generating capacity 1990–2030 (MW)

Type of fuel	1990	1995	2000	2010	2020	2030
Lignite	1,459	2,659	2,625	2,475	1,200	–
Imported coal	–	–	300	9,100	19,800	31,700
Natural gas	4,387	6,864	10,915	10,209	3,296	1,326
Steam – domestic	2,430	3,630	3,630	2,300	–	–
Steam – import	–	–	–	1,400	1,400	–
Combined cycle – domestic	1,831	3,094	4,345	2,509	1,326	1,326
Combined cycle – import	–	–	2,800	4,000	1,200	–
Gas turbine	126	140	140	–	–	–
Oil steam	1,016	1,078	938	5,100	10,800	15,500
LNG	–	–	–	4,800	21,500	39,200
Diesel	522	1,327	1,246	1,546	2,100	2,800
Gas turbine	98	816	746	1,046	1,600	2,300
Diesel engine	424	511	500	500	500	500
Hydro	2,274	2,883	3,383	5,058	4,610	2,393
Total	9,657	14,810	19,406	38,287	63,935	92,918
Imported from Laos	–	–	1,311	1,611	–	–
Interconnection/other	–	–	300	300	–	–
Grand total	9,657	14,810	21,017	40,198	63,935	92,918

Note: totals may not add up due to rounding

Source: Electricity Generating Board of Thailand, in TEI (1995:2–9)

Building renewable energy in grid-dominated areas

Table 6.4: Thailand: planned capacity additions 1997–2011 (MW)

Year	Hydro	Steam		Combined cycle		Gas turbine		Annual total			Grand total
		Total	IPP share	Total	IPP share	Total	IPP share	Total Egat	Total IPP	Purch-ased	
1997	–	–	–	1,029	–	–	–	1,029	–	256	1,285
1998	–	–	–	400	–	–	–	400	–	1,632	2,032
1999	10	700	–	1,900	700	–	–	1,910	700	1,035	3,645
2000	500	975	–	1,400	1,000	–	–	1,875	1,000	–	2,875
2001	–	2,550	1,750	673	673	–	–	800	2,423	–	3,223
1997–2001	510	4,225	1,750	5,402	2,373	–	–	6,015	4,123	2,923	13,060
2002	–	850	350	–	–	–	–	500	350	608	1,458
2003	500	1,341	1,341	300	–	–	–	800	1,341	700	2,841
2004	–	–	–	–	–	–	–	–	–	1,400	1,400
2005	660	690	1,000	300	300	–	–	350	1,300	–	1,650
2006	–	2,690	1,000	–	–	–	–	1,690	1,000	–	2,690
2002–2006	1,160	5,571	3,691	600	300	–	–	3,340	3,991	2,708	10,039
2007	400	2,300	2,300	–381	–	–	–	19	2,300	–	2,319
2008	400	1,990	2,300	220	600	200	200	–290	3,100	–	2,810
2009	–	2,613	3,000	–	–	–	–	–388	3,000	–	2,613
2010	–	2,300	2,300	600	600	400	400		3,300	–	3,300
2011	–	2,225	2,300	–	–	400	400	–75	2,700	–	2,625
2007–2011	800	11,428	12,200	439	1,200	1,000	1,000	–734	14,400	–	13,667
Total 1997–2011	2,470	21,224	17,641	6,441	3,873	1,000	1,000	8,621	22,514	5,631	36,766

Note: Purchases = from Laos and Small Power Producers (SPPs)

Source: Electricity Generating Authority of Thailand (Egat) Power Development Plan (PDP 97-01), April 1997 (in Lefevre et al, 1997b:58)

Thailand's national grid system is highly extensive. Table 6.5 indicates the percentage of households and villages with access to grid electricity in both the central Bangkok (metropolitan) area, and the provinces. This shows that in 1990, some 92.7 per cent of provincial villages and 83.4 per cent of households in the franchise of the Metropolitan Electricity Authority (MEA) (centring on Bangkok) had access to grid-supplied electricity. This high degree of electrification is partly a result of the organizational efficiency of Egat; the fact that Thailand is composed mainly of one large land mass rather than islands and mountains; plus also government policy to help avoid insurgency in remote zones by providing rural electricity to ethnic groups practising shifting cultivation as a way to encourage sedentary agriculture. Electricity is transmitted via 500 kV, 210 kV and 115 kV lines and substations. A variety of new lines and substations are still planned in order to increase the accessibility of the grid system, and to make way for the new power plants designed under the PDP.

Table 6.5: Electrification in Thailand

Authority	1985	1986	1987	1988	1989	1990
Metropolitan Electricity Authority (Bangkok)						
Total number of households (millions)	1.16	1.18	1.28	1.38	1.44	1.53
Percent of households electrified	83.2	84.8	82.0	81.1	82.6	83.4
Provincial Electricity Authority (Regions)						
Total number of villages (thousands)	56.1	57.0	57.4	58.4	59.1	60.2
Percent of villages electrified	69.7	72.3	81.5	83.8	88.7	92.7

Source: adapted from Green, 1997:108

The electricity generation and supply industry in Thailand is therefore highly centralized and dependent on the grid system. The institutions governing electricity have also historically been centralized, although government plans are to liberalize the current structure of organizations. The most important institution governing electricity generation and supply in Thailand is Egat. Egat was formed in 1969 under the Egat Act, which saw the integration of three independent utilities: the Yanhee Electricity Authority, the Lignite Authority and the North-east Electricity Authority. The Yanhee Authority was formed in 1957 for 35 central and northern provinces, the Lignite Authority was also established in 1957 for coal-fired electricity generation, and the North-east Authority was formed in 1962.

Egat has been reformed several times following amendments to the Egat Act in 1978, 1984, 1987 and 1992. The 1992 amendments reiterated Egat's role as the primary generator of electricity, which it can then transmit or distribute to the Metropolitan and Provincial Electricity Authorities. In 1997, some 98 per cent of Egat's sales went to these authorities, with the rest going directly to end-user customers. Egat has a reputation in South-east Asia for being efficient and well organized (Lefevre *et al*, 1997b:40).

The MEA is the chief distributor within the Bangkok region, and the industrial provinces of Samut Prakan and Nonthaburi. The MEA was established in 1958 by the merger of the Bangkok Electric Works and the Electric Division of the Public Works Department. Until 1961, it was responsible for generation as well as distribution, until the generation work was transferred to the Yanhee Authority. The Provincial Electricity Authority (PEA) is responsible for all electricity distribution in Thailand except for the MEA area. It was established in 1960 out of the Provincial Electricity Organization, which itself was created in 1954. Cross subsidies exist between the MEA and PEA in order to provide cheap rates to customers using less than 150 kWh per month (the so-called 'lifeline rate') (Lefevre *et al*, 1997b:50).

In addition to these specifically electricity-oriented organizations, the Petroleum Authority of Thailand (PTT) is also influential by being the main supplier of oil and gas to Egat. In the year to the end of September 1996, PTT supplied some 30 per cent and 28 per cent of Egat's oil and natural gas (Lefevre *et al*, 1997b:45).

Energy and electricity regulation is carried out by two main bodies established in 1986. The most important regulatory body is the National Energy Policy Council (NEPC). This is a senior governmental body chaired by the Prime Minister, and includes representatives from several important ministries such as finance. The functions of the NEPC include setting long-term strategy for electricity generation, energy conservation and alternative energy development. The second regulatory body is the National Energy Policy Office (NEPO). Under the NEPC Act of 1992, the NEPO's role is to act as a secretariat to the NEPC through studying energy trends and options for the NEPC and monitoring and implementing policy decisions. Under the same act, the NEPC and the NEPO act together to approve tariffs set by Egat, monitor PPAs arranged by Egat and undertake general regulatory activity for the electricity supply industry.

Privatization and liberalization

Thailand has undertaken a new programme of privatization in order to increase the supply of new power projects and to improve efficiency of operation. Some 60 per cent of new generating capacity between 1996 and 2011 (22,514 MW) is planned to come from IPPs (see Table 6.4). However, reforms have sometimes been resisted by the bureaucracies that have characterized Thai government, including military government, in recent decades (see Dhiratayakinant, 1989).

Early attempts to privatize state-operated enterprises were proposed in the 1980s, but resisted by the trade unions. Guidelines for privatization were eventually laid out in the sixth National and Economic Social Development Plan for Thailand between 1987 and 1991. The committee created under the Plan announced a strategy in 1988 to privatize some 41 out of 61 state-operated enterprises by 2001. One particular way for this to proceed was through corporatization – or the treatment of state-operated enterprises on profits and losses on the same basis of a commercial operation, even though it remains in state hands. The initial focus for energy was on gas and oil, and in 1991 the PTT was deregulated, and split into four units to deal separately with oil, gas, petrochemicals and general services.

In 1992, the Royal Act on Private (Sector) Participation in State Affairs was promulgated to ensure an open and transparent process for privatization. This law required that any new private-sector projects of over Baht (Bt) 1 billion should be evaluated by the National Economic and Social Development Board, and those with existing assets over Bt 1 billion by the Ministry of Finance before being submitted to the Cabinet for approval. For smaller projects of more than Bt 5 million, there should be an approval process by a consultant accredited by the Ministry of Finance. In 1992, the Committee for Consideration of Increasing Private Sector's Role in Cooperation to Develop State Enterprise was also set up as a further body to identify and monitor privatization projects.

Building renewable energy in grid-dominated areas

Fifteen state-operated enterprises were identified as urgently needing privatization. These included the Bangkok Mass Transit Authority, the PTT, Egat, MEA, PEA and State Railways of Thailand.

Two important reforms took privatization of electricity further. The aforementioned amendment to the Egat Act in May 1992 allowed Egat to create a subsidiary: the Electricity Generating Company Ltd (Egco). This was a holding company that could make its own subsidiaries generating electricity and selling it to Egat through long-term contracts. In March 1994, Egco was listed on the stock exchange, and shares have sold successfully since. The creation of Egco was an attempt to improve commercial efficiency in the generation of electricity and operation of Egat.

The second reform was the introduction of the so-called Four Step Plan by the Cabinet Resolution of 12 September 1992. The steps were:

- Step 1: 1992–3: create better efficiency within Egat through reforms; seek to establish uniform bases for national tariffs; achieve PPAs between Egat, the MEA and PEA; and the establishment of Egco in order to accelerate the privatization and decentralization of Egat.
- Step 2: 1993–4: the subsidiary Egco would proceed to buy the Khanom power plants (in Nakhorn Si Thammarat province) from Egat; Egat itself would be decentralized into smaller administrative units; tariffs would be revised regionally in order to reflect marginal costs; the PEA would also be divided into cost centres.
- Step 3: 1994–5: the corporatization of Egat, PEA and MEA; the privatization of Egat and listing on the stock exchange; and the solicitation of IPPs for power production 1995-2001.
- Step 4: 1995–6: the separation of PEA into regional distribution companies according to established business units; plus increasing Egat's capital and sales to the stock exchange while keeping the government as the main stockholder.

This plan was based upon three aspects of privatization: firstly corporatization in order to increase the commercial efficiency of the state-operated enterprises; secondly the sale of generating assets to the private sector; and thirdly the encouragement of new IPPs to build, operate and own new capacity. Ultimate control over the electricity sector, however, would remain in the hands of government. The plan also used an evaluation of corporatization in which organizations became known as a 'good', or 'excellent' state enterprise if they remitted at least 30 per cent of net profit to the government; they made profit at a rate of at least 6 per cent (replacement cost); they kept labour costs at below 20 per cent of all costs (or 10 per cent for capital intensive industries); and increased productivity annually at 2 per cent (Lefevre et al, 1997b:106). The plan therefore aimed high in achieving both commercial efficiency and increased investment in power generation by the private sector.

The success of the Four Step Plan, however, has not been complete. In particular, full privatization has encountered resistance from within the state enterprises to relinquishing full control of their previous activities. Egat, for example, received its 'good state enterprise' status in August 1994, and has acheived benefits including increased efficiency, greater independence for some parts of its generation activities and a decreased outflow of skilled workers because of an improved salary structure. However, Egat has also asked for a deferral for full privatization until the late 1990s, and has also requested that it not be dissolved totally into a limited company but instead be allowed to create subsidiaries that would operate under its umbrella organization. The reasons for these requests are that full dissolution would require another lengthy amendment to the Egat Act, and that a single business unit was needed in order to maintain efficiency with hydropower when negotiating and planning with other government agencies such as irrigation or forestry. There have also been delays in establishing uniform tariffs.

The PEA has been divided into four distinct regions of north, central (excluding Bangkok), north-east and south, with the aims of transforming each into subsidiary companies. In addition, the PEA may create new companies out of its existing services such as pole manufacturing and construction. However, the PEA has also requested not to be corporatized or to be privatized fully. The MEA has conducted similar reforms in rationalizing costs and identifying businesses within its existing structure.

The problems of implementing privatization have been acknowledged by government and plans adjusted accordingly. The Cabinet Resolution of 11 April 1995 created a new Subcommittee on the Future Structure of the Electricity Supply Industry. Two new phases were identified: 1996–9, allowing full commercialization and selling of shares of Egat, MEA and PEA; and 2000–05, during which the state enterprises would finally be fully privatized by transferring them into limited companies on the stock exchange. The ultimate aim of the reforms would be to create many producers of electricity but one common carrier in charge of transmission. The distributors would remain as the PEA and the MEA, although these would now be decentralized into several separate businesses. The NEPC and the NEPO would remain the main regulatory bodies (Lefevre *et al*, 1997b:98–106).

These reforms have created a platform for liberalizing the existing structures of the electricity supply industry in Thailand. In addition, there have also been reforms to encourage private investment in the power sector. The terms for IPPs were approved by the Cabinet in May 1994, and the first solicitation for IPP projects in December 1994. The terms required that IPP proposals matched Egat criteria for suitability of projects and grid connection, and that they had guaranteed fuel supply contracts and environmental impact assessments. The first solicitation sought to provide 1,000 MW between 1998–2000, and 2,800 MW between 2001–02. This was later increased to an overall total of 5,800 MW. At the closing date of the first solicitation, there were 32 bidders with 50 proposals, totalling 39,067 MW, of which 62 per cent were natural gas, 35 per cent were coal, and the remainder were orimulsion fuel (Lefevre *et al*, 1997b:73–4). A second solicitation for IPPs is scheduled for 1999, seeking to provide 5,600 MW for 2006–08.

The incentives for IPPs are very attractive. The government has placed no restriction on the type of fuel or location chosen, subject to the usual environmental and planning standards. The attraction for foreign investors is underlined by the regulation that subsidiaries of Egat can only make bids for projects when Egat owns less than 20 per cent of the project, and for a maximum of 1,400 MW.

Small Power Producers (SPPs) were introduced as a category of IPP in 1992. There are two broad types of SPP: cogeneration manufacturers and renewable energy producers. These are discussed in more detail in the next section.

In addition to these specific reforms to electricity generation, during the early 1990s Thailand also passed other legislation that impacted on environmental and energy efficiency. The 1992 Environmental Quality Act contained a number of provisions that strengthened the political infrastructure of environmental protection within the state bureaucracy. In particular, it merged the Board of Environment with the Ministry of

---

Box 6.1: Demand side management in Thailand

The fifth Five-year Plan (1982–6) marked the beginning of a new phase in the development of energy conservation strategies in Thailand. Since this planning period, all Five-year Plans have included specific measures for energy conservation, such as the Demand Side Management Programme and the Energy Conservation Programme. Early policies included the elimination of subsidies and the creation of information and training centres on the theme of energy conservation.

In 1985, the Energy Conservation Centre of Thailand was established. This then launched several programmes such as the Energy Audit Programme and the Energy Information Dissemination and Training Programme. Every year between 1986 and 1991, the centre conducted energy audits on about a hundred factories to collect data on power usage.

The seventh Five-year Plan (1991–6) saw the completion of the first DSM project in electric power in 1991. The programme rationalized lighting in the residential and commercial sectors, and installed an efficient motors programme in industry. The programme lasted for five years, and aimed to reduce peak demand by 238 MW, saving 1,428 GWh of electricity by 1997. The cost of this programme was an initial budget of US$189 million.

The 1992 Energy Conservation Act defined measures and targets for energy conservation in the industrial and commercial sectors, and established a fund for promoting energy conservation. The fund was established by transferring the revenue on a tax on petroleum products (at 0.07 Baht, or less than 1 US cent per litre), and money was spent to assist designated factories in carrying out the provisions of the Act. The initial allocation to the fund was 1.5 million Baht (US$60 million), with an expected annual capital inflow of 1.5–2 billion Baht (US$60–80 million).

Source: Lefevre and Bui Duy Thanh (1996); TEI (1997)

Building renewable energy in grid-dominated areas 115

Science and Technology to become the Ministry of Science, Technology and Environment. This also created the Pollution Control Department as a specialist agency to address industrial pollution. The Energy Conservation Act, also in 1992, created a fund for improving energy efficiency, which has been used since to improve DSM of electricity use in Thailand. DSM in fact has been developed to a high degree in Thailand in addition to measures for accelerating the adoption of renewable energy (see Box 6.1).

Renewable energy and technology transfer

Thailand may therefore be identified as a country where there are abundant resources for renewable energy in terms of solar radiation and supply of biomass. However, the market demand for renewable energy is relatively small because of a highly developed grid system and a government policy that has sought to make the use of fossil fuels more efficient rather than actively develop renewable or alternative energy sources. Furthermore, the general success of the government in achieving competition and cost reduction in the production of electricity may make it difficult for renewable energy to compete with fossil fuels. Nevertheless, there has been some progress in developing renewable energy in Thailand. This next section identifies these developments, and attempts to highlight what may be learnt from this experience to draw lessons for building renewable energy within these difficult market conditions.

The section is divided into two parts. The first assesses development in direct investment involving renewable energy technology and focuses on point-to-point technology transfer, or the building of renewable energy through investment alone. This investment may be from foreign or domestic sources, but must involve introducing technologies into regions where there has been prior experience of these technologies. The second part focuses on horizontal integration, or the building of institutional capacity for technology adoption, often including local manufacturers and training. The use of these two parts reflects the book's two aims of assessing how to build renewable energy in industrializing countries through investment, and then how this investment may lead to the transfer of environmentally sound technology.

Vertical transfer

There is comparatively little market for direct investment in decentralized renewable energy in Thailand because so much of the country is already attached to grid supplies. In Table 6.5, it was estimated that in 1990, some 92.7 per cent of villages outside Bangkok, and 83.4 per cent of households inside Bangkok had grid-supplied electricity. Direct investment or vertical transfer of renewable energy technology may therefore be limited to those applications that can be connected to the grid, or be used in the small areas still without grid supply.

Government policy has attempted to develop PV technology since 1976. Official bodies such as the Ministry of Public Health and Medical Volunteers Foundation installed PV panels at rural health stations. The Telephone Organization of Thailand also undertook research to indicate which stations might be suitable for PV power. The confusion of the oil shocks in the 1970s, however, dissuaded further action in this area until renewable energy was targeted in the fifth Five-year Plan of 1982–6 (Kirtikara, 1997). Under the fifth plan, the National Electrification Administration (NEA) with USAID brought together ten government agencies to work together on building PV technology. The Telephone Organization of Thailand finally installed PV units at 50 signal repeating stations, and the PEA installed three PV/battery power plants in three rural villages as test projects with assistance from Japanese ODA.

Under the sixth Five-year Plan, the Department of Public Welfare committed itself to installing PV equipment in 50 to 100 villages annually until 2001. The Ministry of Defence also undertook PV installation in north-eastern Thailand under the so-called 'Green Isarn (north-eastern Thailand) project' between 1988–92. However, other aspects of this project such as reforestation of state-owned land where there were thousands of squatter farmers were not successfully carried out, and the Green Isarn project lost credibility with the replacement of the military government after democratic elections in 1992. Since then, the main impetus for PV development has been the Department of Public Welfare and Department of Energy Development and Promotion under the Ministry of Science, Technology and Environment for purposes such as water pumps in arid locations, particularly the north-east. By the end of 1996, about 2.5 MW of PV were installed in Thailand, of which 90 per cent had been funded by the government.

The Energy Conservation Act of 1992 classified new and renewable energy development under the category of voluntary programmes. Renewable energy took second priority behind measures such as energy conservation within government and other designated factories and buildings. However, since the democratic elections of 1988 and 1992, the government has been seeking a greater involvement of international agencies in renewable energy production, and private-sector investment. The Australian International Centre for the Application of Solar Energy (CASE) has been one international agency that has responded to this invitation. By August 1997 CASE had four projects in Thailand seeking to use high-level renewable energy technology to power in both off-grid and grid-supplied locations (see Box 6.2).

The projects undertaken by CASE have attempted to introduce imported microhydro, PV and wind power technology to rural applications in conjunction with diesel generators and battery recharging. The projects have been undertaken in collaboration with the PEA, the Curtis University of Technology in Western Australia and AusAid. To date, work has attempted to demonstrate the potential of renewable power sources in Thailand. However, the projects have sought to indicate how renewable power may be integrated with local grid supplies, or dependency on batteries, rather than attempting to replace either of these.

Horizontal transfer

Horizontal transfer of renewable energy technology may occur through encouraging the adoption of renewable energy as a supplement to grid-supplied electricity, or through the building of the institutional capacity to train and increase the demand for renewable energy. The recent

> **Box 6.2: CASE renewable energy projects in Thailand**
>
> The Australian International Centre for Application of Solar Energy (CASE) had by 1997 established four pilot projects in Thailand to demonstrate the potential of PV, microhydro and wind technology. Two projects are stand-alone hybrid systems, and two are grid connected. The projects demonstrate both vertical transfer of imported technology, plus the horizontal integration into each locality through training and planned power uses. All projects have conducted local training of villagers to ensure maintenance and an understanding of the system potential. All systems also have remote monitoring through telecommunication links.
>
> 1. Ban Khun Pae: a mountain village 120 km south-west of Chiang Mai inhabited by the Karen ethnic minority. There are about 90 households in the village who need electricity for lighting, refrigeration, schools and cooking. The project is using 7.2 kW (6 x 120W) of PV, in array; a pre-existing 90 kW minihydro system; plus a 110 kW battery storage and 40 kW AC/DC inverter.
>
> 2. Ban Den Mai Sung: a lowland village 300 km south of Chiang Mai in Tak province, of some 200 households. A 60 kW PV/battery/inverter power system was commissioned in 1986, which was then connected to a local grid after the grid was extended in 1990. The project aims to integrate the solar input into the grid and reduce the potential power fluctuations.
>
> 3. Ban Mai Ka See: a lowland village 400 km north-west of Bangkok in Nakhon Sawan province. This was fitted with the same power system as Ban Den Mai Sung in 1986, which was then also connected to the grid in 1990. The project is now working to identify ways to increase the efficiency of connecting PV to grid supplies.
>
> 4. Ko Kut Island: a small island 80 km south of Trat, in the Gulf of Thailand close to the Thai–Cambodian border. There are approximately 170 households conducting agriculture and fishing, with a variety of domestic and commercial electricity uses, including a temple, school and health clinic. Existing power came from two 50 kW diesel generators. CASE installed 12 kW of PV in array (100 x 120W); plus 110 kW battery storage and a 40 kW AC/DC inverter. It is also planned to install wind turbines, and the project is measuring wind speeds to facilitate this.
>
> Source: *International Solar News*, August 1997

programme of privatization has achieved this partly in relation through the SPPs.

SPPs were introduced in 1992 as part of the privatization of Egat. SPPs are essentially IPPs that produce small amounts of electricity in the region of between 50 MW and 90 MW, however, in practice they are commonly specialized users of cogeneration, and new and renewable energy technologies.

> **Box 6.3: Examples of Small Power Producers in Thailand**
>
> **1. The Cogeneration Co. Ltd at Map Ta Phut**
> This model of a Small Power Producer was established in 1996 in the Map Ta Phut industrial estate in Rayong province on Thailand's eastern seaboard. It has 300 MW of cogeneration, from two 150 MW units.
>
> The owner is The Cogeneration Co. Ltd, (or 'Coco'), which was formed by a joint venture between Banplu plc and Nordic Power Investment AB. Banplu is a leading coal producer in Thailand, and owns 72 per cent of Coco. Nordic owns the remainder, and is a consortium of four Scandinavian power electric utilities including Elkraft (Denmark), Imatran Voima Oy (Finland) and Sydkraft AB and Vattenfall AB (of Sweden).
>
> The project will sell process steam and electricity to industries in the Map Ta Phut estate, and sell any surplus to Egat. The main contractor in building the power units was GEC Alsthom (of France), with each 150 MW unit constructed separately. Each 150 MW unit sells 90 MW to the grid, the maximum wattage allowed under the SPP legislation.
>
> In December 1995, Coco undertook a public offering of 31 million shares at US$2.31 each. The offer was oversubscribed.
>
> *Source: International Private Power Quarterly (in Lefevre et al, 1997b:86)*
>
> **2. TRT Parawood Co. Ltd, southern Thailand**
> This company is one Thailand's largest rubberwood sawmill operators, with a production of 4,000 m^3 $month^{-1}$. In the late 1980s the company decided to invest in a 2.5 MW cogeneration plant for internal use, and to sell excess power to the grid. Commissioning started in 1995.
>
> The plant consists of an automatic system for silo unloading, fuel distribution and boiler feeding. The boiler produces 21 tons $hour^{-1}$ of superheated steam at 2.6 MPa and 320 °C, and includes a predryer for wet sawdust, a dust collector, and a 2.5 MW extraction condensing turbo-generator. Five tons of process steam at 0.7 MPa are required for kiln drying operations. The fuel used is waste biomass material from the factory.
>
> The total cost of the equipment was estimated to be US$2.2 million, including the civil works and building structures. However, by producing its own power, the company can save US$840,000 per year in reduced energy costs, plus sell excess power to the grid at an estimated US$48,000 per year.
>
> *Source: Green (1997:114)*

The first offer to buy electricity from SPPs came on 1 April 1992. By August 1997, there were officially 20 SPPs supplying the national grid, with a total supply rising from just 13.3 MW in 1994 to 1,215.1 MW in 1996 (Lefevre *et al*, 1997b:85). The first SPP was a sugar mill in Nakhorn Sawan province, some 240 km north of Bangkok. This generated 4.8 MW of electricity using bagasse as fuel. It was connected to the grid using a 22 kV line belonging to the PEA, and was bound by a flexible non-long term sales contract. A further SPP is the Map Ta Phut cogeneration plant in Rayong which sells 180 MW per year to Egat on a firm basis. This plant is generally considered to be an important and influential model for further SPPs – see Box 6.3. By August 1997, there were a further 46 SPP contracts signed with Egat for a potential 1,855 MW.

The number of SPPs is therefore rapidly increasing in Thailand, and these create opportunities for renewable and alternative energy technologies. Table 6.6 shows the growth of SPPs according to generation capacity and contribution to grid. Table 6.7 indicates the number of SPPs according to fuel type, including type of biomass.

In addition to the encouragement of renewable energy through SPPs, the Thai government has also sought to build local capacity for the adoption of renewable energy. The Energy Conservation Act has awarded

Table 6.6: Thailand: projected Small Power Producers as of August 1997

	Firm supply contract[1]	Non-firm supply contract[2]	Total
Proposals submitted			
No. of projects	66	22	88
Generation capacity (MW)	7,615.4	619.7	8,235.7
Sale to Egat (MW)	4,423.5	188.7	4,621.2
Proposals accepted			
No. of projects	38	22	60
Generation capacity (MW)	4,211.4	619.7	4,831.1
Sale to Egat (MW)	2,356.5	188.7	2,545.2
Contracts signed			
No. of projects	25	21	46
Generation capacity (MW)	3,002.4	593.7	3,596.1
Sale to Egat (MW)	1,669.0	185.7	1,854.7
No. supplying power to the grid	4	16	20

Notes: 1: Firm contracts are for more than five years and receive payments for specific capacity, thus specifying how much electricity is demanded.
2: Non-firm contracts are for less than five years, and receive no capacity payments, hence not specifying how much electricity is demanded.

Source: NEPO (in Lefevre et al, 1997b:87)

Table 6.7: Thailand: accepted Small Power Producer proposals by fuel type, as of August 1997

	No. of projects	Generation capacity (MW)	Sale to Egat (MW)
Commercial energy			
Natural gas	25	3,219	1,792
Fuel oil	1	10	9
Coal	10	1,110	556
Total	36	4,339	2,357
Renewable energy			
Bagasse	14	300	65
Paddy husk, wood chips etc.	9	142	82
Palm oil	1	50	42
Total	24	492	189
Grand total	60	4,831	2,546

Source: NEPO (in Lefevre et al, 1997b:87)

funds to projects researching the use of biogas and biomass in rural industries such as curing tobacco leaves. Demonstration projects have also sought to provide training and knowledge about new technologies for further development. Egat, for example has three demonstration projects in Aranya Prathet (300 km north of Bangkok); Phuket Island (on the western coast of the southern peninsula); and in Sankamphaeng near Chiang Mai, which use a combination of imported PV and wind hybrid systems introduced in 1990. The New Energy and Industrial Technology Development Organization (NEDO) of Japan also has two PV demonstration projects in Ban Khao Kang Rieng (200 km northwest of Bangkok) and Tao Tai Island (15 km east of Phuket) (Green, 1997:109–10).

Commercial joint ventures for the local manufacturing of PV are also emerging. BP Solar (of the UK/Australia) and Solartron (of the United States) import fabricated solar cells and undertake lamination and encapsulation in Thailand. The potential for similar joint ventures with wind power technology is limited because of the few locations in Thailand with suitable wind speeds. Nonetheless, efforts to research wind speed and implement wind-powered water pumps in the north-east have been taken by Egat, Chulalongkorn University and the King Mongkut

Institute of Technology in Bangkok. A small private manufacturer, U-Sa Economic Development Co. Ltd has successfully designed and constructed a multi-blade type windmill for water pumping (Woravech, 1997).

Summary and conclusions

The case study of Thailand was looked at in detail because it represented market conditions which might – in theory – make it difficult for investment in renewable energy. These conditions were a high local dependency on grid-supplied electricity, and a government policy that encourages private investment in fossil fuel power production as the cheapest option.

Evidence concerning renewable energy and technology transfer reflects these market conditions. Any investment that has occurred in renewable energy has been to supplement the local grid (as with the example of CASE), or local cogeneration factories producing biomass fuel where again the surplus is also sold to the grid. There has been some limited off-grid investment, but this has been dominated by the need to integrate this into existing local needs such as battery recharging, rather than any attempt to provide decentralized electricity generation as a long-term alternative to grid supply.

Nevertheless, the examples of building renewable energy under such conditions of grid domination indicate some success, and provide a model for similar development elsewhere. The most important step has been the SPP legislation, which has provided a commercial incentive for small electricity producers to develop power cogeneration or renewable energy technology independently of centralized planning by Egat. The SPP legislation provides a model in terms of both technology and pricing of how to integrate renewable energy development into a grid system.

Direct, vertical investment of new renewable energy technology in Thailand is less developed than the SPP scheme. The main barrier to investment in decentralized power equipment such as PV is the large degree of grid connection. However, public organizations such as the

Building renewable energy in grid-dominated areas 123

Telephone Organization of Thailand and the Ministry of Defence, or non-profit making development agencies such as CASE have conducted pilot projects to demonstrate that these technologies can be made to work. In time, these projects may force a wider adoption amongst other government buildings and operations, and build larger market demand. Some joint ventures in manufacturing PV and wind technology are also emerging, which indicates that international investment may be starting to trickle down to local manufacturers.

Privatization and liberalization in Thailand has therefore opened up opportunities for IPPs and this has had the greatest potential for international investment in grid-connected large hydro and small biomass development rather than decentralized, off-grid, systems. The creation of the SPP legislation suggests that it is possible to influence electricity supply towards renewable energy production. However, in Thailand, perhaps the greatest steps in energy and environmental policy may come from DSM rather than renewable energy development because of the dominance of grid and fossil-fuel related power sources.

Chapter 7

Renewable energy investment under dominant state ownership: the case of Vietnam

Introduction

This chapter continues the analysis of different market conditions for renewable energy investment in South-east Asia by focusing on category 2 of Table 5.1. Under these conditions, the opportunity for renewable energy is limited by the existence of a centralized grid supplied by alternative energy sources. Furthermore, foreign investment is made difficult by the existence of a high level of regulation and a low level of private-sector ownership or operation of the electricity sector. Such circumstances suggest that renewable energy investment may be uncompetitive and difficult to achieve. However, at the same time, the strict level of regulation may enable market loopholes to be opened to enable renewable energy investment, or the building of institutional capacity to support the adoption of renewable energy technology.

The case study chosen for this chapter is Vietnam. Vietnam is a valid example because it has few islands in comparison with Indonesia or the Philippines and therefore may achieve electrification through extending a centralized grid from its coastal cities. Furthermore, the electricity sector in Vietnam is still dominated by bureaucratic state-operated enterprises as a result of its ongoing transition from communist economic and political management to a market-oriented economy. Private ownership and private-sector involvement in power generation are still poorly developed. There are also large indigenous supplies of coal and gas in Vietnam, which present a challenge to the development of renewable energy sources.

Yet against this, Vietnam also has large territories still unconnected to the national grid. The country still has an interventionist government that may wish to accelerate economic development in rural areas in order to advance public welfare and avoid the macroeconomic drain caused by rural–urban migration under conditions of rapid industrialization and economic transition. The electricity sector may therefore be characterized by strict government control, but within this may be the opportunity for harnessing new technologies and integrating these horizontally into rural development schemes.

The chapter starts by summarizing economic and energy trends in Vietnam before discussing in more detail recent measures for privatization and liberalization. The impacts of these on renewable energy investment and technology transfer are then assessed before drawing conclusions about which institutional structures have enabled the greatest success in increasing the adoption of renewable energy.

Energy supply and renewable energy resources in Vietnam

Overview

Vietnam is located on the far eastern coast of mainland South-east Asia where it is composed of one large mainland and several small islands. Its total land surface area is 327,500 km^2, and it has a vast coastline from China to Cambodia of some 2,500 km. Three-quarters of the country is hilly or mountainous, particularly in the far north and north-west, where there are remote settlements inhabited by a variety of ethnic groups including shifting cultivators. In the far south, the landscape is characterized by the Mekong River floodplain, from where most agricultural production comes. In between north and south, the country narrows to just 50 km at one point in-between the South China Sea and the neighbouring country of Laos (see Figure 7.1).

Vietnam's climate is monsoonal and varied between regions. Generally, from November to April the north experiences a relatively cold and humid winter followed by a rainy summer in which typhoons are experienced. The average temperatures for the north are 20–27 °C. In the south and central parts of the country, the cold periods are shorter, and there is a more clearly defined rainy season from May to October. Average temperatures are 23–27 °C. The average annual rainfall for Vietnam is 1,500mm.

The population of Vietnam was estimated in 1993 to be 72 million. Most citizens conduct agricultural lifestyles, and live on the low-lying land close to the coast and the Mekong Delta. The urban population has been estimated by the ADB to be some 30 per cent of the total. However, accurate statistics are difficult to establish especially as rural–urban migration has been rapid in recent years, and that many rural workers migrate to cities for short periods each year to find work. Within the more remote and mountainous zones of the country close to the Laotian and Chinese borders, there are also many ethnic minorities and agricultural communities that are not integrated into the national economy and live semi-autonomously from the lowland Vietnamese and state.

Vietnam became unified under communism in 1975, but by the late 1980s was already passing legislation to become more market oriented. This programme, known as 'Doi Moi' (or 'renovation') is still in progress, but has gradually led to the abolition of state subsidies and controls on pricing and supply in favour of using market forces to allocate resources. These changes became more formal in 1989 with the devaluation of the Vietnamese currency, the Dong, and in 1994 when the United States ended its trade embargo. In 1995 Vietnam was admitted to the Association of South-east Asian Nations (ASEAN) and the United States normalized diplomatic ties. Since then, foreign investment has increased rapidly, and GDP grew at 5.9 per cent between 1985 and 1990, and then at more than 8 per cent per year during the early 1990s – a rate higher than all other countries of South-east Asia and even China. In 1994, Vietnam's GDP totalled US$16 billion, of which 75 per cent came from oil and rice exports.

However, this growth has been achieved at the cost of immense social and economic upheaval resulting from the dropping of subsidies for small state-operated enterprises and cooperatives, and the rapid growth of cities and industrial zones. Most industrial growth has occurred in the southern provinces of Ho Chi Minh City and Dong Nai, rather than near the capital city of Hanoi in the north. Connected to this industrial growth, the position of the ruling Vietnamese communist party is also changing following the involvement of many senior and non-senior cadres in new business activities, and the increasing public challenge to what many see as a betrayal of government principles through corruption.

These social and political changes are impacting on the ability of government to enact political reforms in Vietnam, and also on the trust foreign investors have in joint ventures or local authorities. This is also reflected within the internal organization of Vietnam where some provincial governments have deliberately ignored the instructions of Hanoi by setting up industrial estates of their own in order to encourage local development, rather than letting the central government implement a national strategy. As a result of these legislative problems, the common image of Vietnam as a strictly regulated investment regime may in fact be false because so many of the command-and-control measures are difficult to enforce. When these problems are combined with the financial downturn recently experienced as a result of the South-east Asian financial crisis, the political and economic situation in Vietnam is less promising than it seemed in the early 1990s.

Energy trends

Vietnam is relatively well endowed with coal reserves, and has huge potential for offshore oil and gas development on account of its proximity to the South China Sea. The first oil reserves were discovered in 1975, and in the late 1990s, the country's proven oil reserves were estimated to range between 1.5 and 3 billion barrels, similar to the potential of Australia and Malaysia (although with a less impressive per capita ratio). Proven reserves of gas are estimated to be 3.6 trillion cubic feet, of which 2.3 trillion is associated with oil. The government of Vietnam aims to produce 20 million tonnes (or 400,000 barrels a day) by 2000, but this is still relatively small compared with Indonesia's 1.3 million barrels a day production (Harvie, 1997:59).

Coal production expanded from 4.6 million tonnes in 1990 to 5.9 million tonnes in 1995, and crude oil production increased from 40,000 tonnes in 1986 to 7.6 million tonnes in 1995. Vietnam has a variety of reserves, ranging from an estimated 2,258 million tonnes of low sulphuric anthracite in the northern province of Quang Ninh to 30,000 million tonnes of lignite in the Hog River Delta. In addition to these, Vietnam also has some 150 million tonnes of bituminous coal and 1,000 million tonnes of peat (Breeze, 1996:158). In 1998, Vietnam exported 2.7 million tonnes of coal.

Despite this increase in production, Vietnam has one of the lowest levels of commercial energy consumption in the developing world, with less than 100 kg of oil equivalent per capita per year in 1990 compared with 246 kg in the Philippines, 468 kg in Thailand and 954 kg in Malaysia. The total use of all primary energy in 1990 was 17 million tonnes of oil equivalent. Of this, 61 per cent came from traditional sources such as fuelwood. For commercial energy use, petroleum products supplied 18 per cent, coal 13 per cent and hydroelectricity 8 per cent (Harvie, 1997:58). However, this production is unlikely to satisfy future energy demands within Vietnam, forcing the government to focus further on increasing the speed and efficiency at which it utilizes indigenous resources, as well as seeking cheap energy imports.

Renewable energy resources

Vietnam has a variety of renewable energy resources ranging from large and small hydro plus small-scale solar, wind and biomass applications. However, accurate figures on potential resources are rare. Total theoretical hydropower potential has been placed at 30,000 GWh pa, of which 181,000 GWh are in the north, 89,000 GWh are in the central provinces and 3,000 GWh are in the south. However, of this, only 82,000 GWh are economically feasible for development, and this equates to 15,600 MW of installed capacity (Breeze, 1996:157). By 1997, about 3,000 MW of this had been developed and 800 MW are under construction in large dam projects. In addition, some 10 per cent of hydropower's potential is estimated to be available from small hydropower projects. Currently small and microhydropower units are estimated to contribute 65–120 million kWh for residential use, agricultural production and small industrial requirements in mountainous and midland and high areas (Toan, 1997).

Average solar insolation per month in Ho Chi Minh City is 4.72 kWh m^{-2}, and in Hanoi, 4.08 kWh m^{-2} (PA Energy, 1997). In the south and centre of Vietnam, the insolation remains broadly the same throughout the year, but in northern Vietnam it may fall to as low as 2.40 kWh m^{-2}, therefore lowering its potential contribution to PV electricity. It has also been estimated that Vietnam has 10 million hectares of productive forest, which supplies 21,000 million tonnes of fuelwood a year (Breeze, 1996:158).

Structure and liberalization of the electricity supply industry

Electricity market and institutions

The Vietnamese electricity supply sector at present is characterized by a generally low level of installed generating capacity and a high dependency on hydroelectric power. Table 7.1 indicates the capacity in 1995, of which some 63 per cent came from large dam projects such as at Hoa Binh, which was constructed during the 1970s–1980s with the assistance of Soviet aid. With a capacity of 1,920 MW, Hoa Binh is the largest operating hydropower plant in South-east Asia. A further series of new power plants is planned at Phuy My at Ba Ria in Vung Tau province in southern Vietnam. These projects, partly assisted by the World Bank, are planning to use natural gas piped ashore from the South China Sea and could add 1,650 MW to national installed capacity. Total national installed capacity is planned to be 8,300 MW in 2005

Table 7.1: Vietnam: installed electricity generating capacity, 1995

Types of power plant	Capacity (MW)	Percentage
Coal fired thermal	645	14.4
Oil fired thermal	198	4.4
Gas turbine	384	8.5
Diesel	434	9.7
Hydropower	2,824	63.0
Total	4,485	100.0

Source: Toan (1997:3)

(Breeze, 1996:159-61). Table 7.2 indicates the annual consumption of electricity in Vietnam by sector from 1981–94.

The Vietnamese electricity sector is consequently at an early stage of development when compared with other South-east Asian countries such as Thailand or even Indonesia. Table 7.3 illustrates measured growth in per capita electricity consumption from just 92 kWh per capita in 1986 to 168 kWh per capita in 1994. However, for large areas of rural Vietnam current levels of consumption are likely to have remained unchanged over this time period. Indeed, rural electricity consumption has been estimated to be an average 74 kWh per capita (Quinn, 1997:54).

Table 7.2: Vietnam: annual consumption by sector, 1981-1994 (GWh)

	1981	1985	1990	1992	1994
Residential – commercial	1,030	1,409	2,035	2,153	3,895
Industrial	1,659	2,113	2,846	3.192	3,848
Agricultural	256	308	587	975	1,418
Other	37	36	718	615	–
System losses	954	1,203	2,548	2,716	3,034
System losses (%)	25.8	23.7	29.2	28.1	24.7

Source: ADB, 1994; Institute of Energy, Hanoi, 1995 (in Breeze, 1996:164)

Table 7.3: Vietnam: growth in per capita electricity consumption 1986–1994

Year	1986	1987	1988	1989	1990	1991	1992	1993	1994
Per capita consumption kWh pc pa	92	98	107	120	131	134	138	150	168

Source: Toan (1997:4)

Table 7.4 indicates the projected increase in demand for power in Vietnam until 2010. This table is based on a prediction that demand will increase by between 11.6 to 13.7 per cent annually between 1995 and 2010, and reflects the rate of growth since 1985 when demand was some 5.4 GWh. It has been estimated that in 2010, the proportion of energy produced by hydropower will decrease to 50 per cent of the total, coal will increase slightly to 18 per cent, and gas will produce in the region of 20 to 30 per cent (Ministry of Industry, in *Vietnam News* 31 October 1997:5).

Table 7.4: Vietnam: electricity power demand forecast up to 2010

Year	Low case		Base case		High case	
	GWh	MW	GWh	MW	GWh	MW
1995	14,500	2,800	14,500	2,800	14,500	2,800
2000	17,400	4,900	30,000	5,400	33,000	5,900
2005	45,500	7,750	53,700	9,200	61,200	10,400
2010	75,200	12,300	87,600	14,300	99,100	16,200
Growth per year	11.6%		12.7%		13.7%	

Source: Toan (1997:4)

The transmission and distribution of power is also undergoing rapid growth. In 1993, the government completed a 1,400 km 500 kV transmission network in order to unify three separate utilities and create a single grid. By 1996, the grid system included more than 72,000 km of line and 19,000 Mega Volt Amps (MVA): this represented an annual growth 1994–6 of 9.5 per cent for transmission lines and 13.9 per cent for substations. In January 1997, the government claimed that the national grid system included 98.1 per cent of cities and towns; 88.8 per cent of districts; and 60.2 per cent of villages in Vietnam (Ministry of Industry, in *Vietnam News* 31 October 1997:5). However, estimates by the ADB concerning the total population suggest these figures are exaggerated (Breeze, 1996:162).

Future development of the grid by 2005 is planned to include an additional 800 km of 500 kV lines (plus 2,600 MVA); a larger expansion of some 5,000 km of 220 kV (plus 14,000 MVA); and a major growth in small grid lines of 110 kV by 150,000 km (9,000 MVA) for rural areas. However, the likelihood of achieving these aims is small given the shortage of funds and general problems of investment in Asian infrastructure following the financial crisis after 1996. Furthermore, most transmission systems in cities date from colonial administrations before the Second World War, and are highly inefficient and needing maintenance.

Vietnam's institutions for managing its electricity supply industry have been created and reformed since the 1990s, and consequently it is simpler to describe the institutions at the same time as measures for privatization.

Privatization and liberalization

Vietnam has been undertaking a comprehensive programme of economic liberalization and market orientation since the start of 'Doi Moi' in 1986. The reforms to the energy sector have formed part of these, and must be seen as reflective of these general trends rather than as approaches specifically for power production. In the case of many major industries, including energy, the trend towards privatization has coincided with the creation of organizations by the state to manage these industries for the state. In this sense, the privatization and liberalization of energy in Vietnam has also marked the enforcement of state control rather than the liberalization of existing institutions such as in Thailand (Yuen *et al*, 1996).

The first major law affecting the energy sector was the Petroleum Law of 1993. Vietnam had experienced rapid growth in its oil and gas sectors during the early 1990s as a result of the influx of foreign investors in exploration. The law sought to increase Vietnam's attractiveness to foreign investors and to ease investment. Some of its key provisions included that the profits rate tax for exploration and production be fixed at 50 per cent; and that the rates of oil production royalties would range from 6 to 25 per cent and those for natural gas from 0 to 10 per cent. The purpose of this legislation was to define the potential costs more clearly for investors, and to make terms easier than in some competing countries.

Reforms in the power sector began in December 1994. Changes to the electricity industry reflected those occurring elsewhere in activities such as cement, paper, banking, coal, steel and textiles. Steps taken involved creating large industry monopolies in order to reduce bureaucracy plus also set in motion large new conglomerates that might compete overseas (similar to the *chaebol* of South Korea) (Quinn, 1997). Before these reforms, industrial activities such as power generation were conducted by relevant ministries as part of central planning by government.

Electricity of Vietnam (EVN) was created in early 1995 through the merger of the country's three regional power companies (PC1, PC2 and PC3, and centred on the three main urban centres of Hanoi–Haiphong, Da Nang–Hue, and Ho Chi Minh City and the Mekong Delta). The grids operated by each of these were eventually linked in 1993 by the completion of the aforementioned new transmission line. Under its 1995 constitution, EVN had three main responsibilities: to carry out the state electricity plan concerning investment in generation, transmission and distribution, including creating joint ventures; to organize human resources in the electricity sector; and to receive and develop funding for these projects.

At the same time as EVN was created, the old Ministry of Energy was dissolved, and EVN was made responsible to the Ministry of Industry. The Electricity Department of the Ministry of Industry is EVN's main regulator, and approves all pricing and capital investment decisions. EVN itself was divided into 12 separate cost centres for each power plant, and four transmission companies (dependent accounting units) organized on a regional basis and four construction companies.

Yet despite these initial steps for increasing efficiency, there still has been cross subsidization between different end users, and a desire by government to subsidize electricity expansion at below cost price. In 1990, the World Bank estimated that the tariff system employed was equivalent to only 60 per cent of the long-run marginal cost of supply. Tariffs were revised in 1992 in order to increase EVN revenue, but the average tariff by 1995 was 4 US cents per kWh while the long-run marginal cost was 7 cents per kWh (Breeze, 1996:164).

Furthermore, there has generally been an insufficient framework for guiding the integration of IPPs into the Vietnamese electricity industry. A planned electricity law for international investment is planned for the late 1990s, and in the meantime EVN has published a 'grid code' to provide the guidelines for private-sector participation. At present, IPPs are governed by the 1993 BOT Law and its implementing regulations. This legislation was one of the first important reforms to introduce foreign investment in the provision of infrastructure.

Progress towards private-sector involvement, nonetheless, remains slow. The EVN has stated that it intends IPPs to contribute some 30 per cent of new capacity ($300 million per year). However, by mid 1997 only one IPP had signed a PPA: the Hiep Phuoc Power Company, in March 1997, for a period of seven years to EVN. To help create more private-sector interest, EVN has allowed IPPs inside Industrial Zones to sell power directly to industrial customers at prices negotiated between the IPP and the customer, without reference to EVN. If such arrangements could be extended to customers outside Industrial Zones then IPP participation would increase.

BOT projects have also been proposed to EVN by companies including Oxbow and Enron (of the United States) and BHP (of Australia) but have been complicated by long negotiations and uncertainty on details of PPAs. However, in the longer term, IPP participation may also be threatened by the need to receive supplies of fuel from state agencies PetroVietnam (oil and gas) and Vinacoal (coal), which are not as advanced in privatization as EVN. Furthermore, the ruling communist party has reiterated at party conferences that private-sector participation in crucial industries like power is still ideologically sensitive, and that ultimate control will still lie in state hands (Birchall, 1997; Quinn, 1997).

The progress of privatization in general is also threatened by a loss of confidence in Vietnam from foreign investors on account of the difficulties encountered in making projects work. Typically, investors who wish to invest outside 100 per cent foreign-owned factories in industrial estates are asked to undertake costly joint ventures with indigenous Vietnamese companies in order to maximize the transfer technology and manufacturing standards. In return for this sharing of information, foreign investors often receive access to well-located sites or proximity to markets.

Increasingly, however, investors are resisting this pattern. For example the US toiletries manufacturer, Proctor and Gamble petitioned the Vietnamese Prime Minister in 1997 to dissolve its joint venture and increase capital expenditure in order to avoid the bankruptcy of its Vietnamese subsidiary. Proctor and Gamble claimed that the joint venture has been required to spend too much on administration, overstaffing and misplaced advertising campaigns, resulting in a loss of some US$28 million after just two years of operation (*Vietnam Courier*, 2–8 November 1997, p7). Similarly, in May 1998, the Coca Cola company requested that its Vietnamese subsidiary and joint venture (Coca Cola Chuong Duong) become 100 per cent foreign owned after losing a reported US$12.4 million (*Vietnam News*, 22 May, 1998). In both cases, the US-based investors sought to avoid the costs of joint ventures by becoming totally foreign controlled. These examples suggest that Vietnam is experiencing problems in establishing successful business regulation which allows both foreign investors and the local government to benefit.

Renewable energy and technology transfer

Vietnam may therefore be defined as a country undergoing rapid industrialization and reform of its public-sector utilities. Private-sector involvement in energy provision is still poorly developed, but will be needed in order to ensure that the country's growing demand for power is satisfied. Furthermore, decentralized provision of electricity may enable a faster provision of power to remote rural areas which are still unconnected to the grid system.

This next section assesses the potential for integrating foreign investment with renewable energy development. The section is divided into vertical and horizontal forms of investment and technology transfer. However, as there is still little totally free direct investment from foreign companies, much point-to-point relocation of technology is carried out in conjunction with ODA and domestic development programmes. Some vertical integration may also take place with Vietnamese producers selling new technologies to remote regions.

Vertical transfer

It is difficult to differentiate between vertical and horizontal technology transfer in Vietnam because few international investment projects exist, and most projects are integrated somehow into developmental programmes of the Vietnamese government or international development agencies. To overcome this blending of the two categories, this chapter divides vertical and horizontal technology transfer into firstly those projects that involve implementing new or imported technology into regions in a commercial manner, and secondly, actions to increase local capacity for renewable energy that may not involve importing new technology. However, both of these categories involve aspects of vertical and horizontal integration.

Most attention on building new renewable energy technology businesses, from a vertical integration viewpoint, in Vietnam has been focused on large hydropower projects. Table 7.5 lists existing and proposed hydropower facilities in Vietnam as of May 1996. These projects may be subject to local protests given that they will inevitably mean relocating villages.

Some of these projects have created opportunities for foreign investors. The Thac Mo 150 MW project has been implemented with Ukrainian assistance and equipment, reflecting Vietnam's old trading links with the former Soviet Union. The Vinh Soc 66 MW project was the first major project with Western involvement, with assistance from Electricité de France in 1990. A French consortium led by Cegelec also supplied equipment. Japanese consultants have assisted with the Ham Thuan 300 MW project (Head, 1994).

However, there has also been recent interest in building smaller renewable energy applications. One project co-organized by the Vietnamese Women's Union (VWU) and the Solar Electric Light Fund (SELF) has sought to introduce US-manufactured PV and battery technology into rural villages, but in conjunction with local organizations. The project also aims to recover investors' costs by seeking payment from end users (see Box 7.1).

The initial funding for this project was provided by SELF, the Rockefeller Brothers Fund and Sandia National Laboratories in the United

Table 7.5: Vietnam: existing and proposed hydropower facilities as of May 1996

Existing projects			Proposed projects		
Project	River basin	Installed capacity (MW)	Project	River basin	Installed capacity (MW)
Hoa Binh	Da/Red	1,920	Yali	Srepok	700
Trac Ba	Red River	108	Plie Krong	Se San	123
Vinh Soc I	Con	66	Can Don	Dong Nai	60
Thac Mo	Be	150	Ham Thuan	Dong Nai	34
Song Hinh	Ba	66	Dami	Dong Nai	36
Da Nhim	Dong Nai	160	Son La	Da/Red	2,400
Tri An	Dong Nai	420	Dong Nai 4	Dong Nai	64
			Ban Mai	Ca	347

Source: World Bank et al (1996:37)

Box 7.1: The SELF – Vietnamese Women's Union Solar Energy Project

The project is entitled,'Solar Project in Support of Rural Women and Children', and began in December 1994 in the three provinces of Tra Vinh and Tien Gang in the south, and Hoa Binh in the north. The project provided PV technology to five communes, in the first stage a total of 130 Solar Home Systems (SHS) were installed, including two street lighting systems.

The SHS were manufactured by United Solar Systems Corporation of the United States. The project provided training to ten technicians and also involved monitoring the success of the equipment and the availability of maintenance facilities if necessary. The capacity of the SHS were 22.5 Wp each.

Households paying for the SHS make an initial down-payment of 20 per cent, and then pay monthly contributions for four years. The second phase of the project is to increase the module size to 35 Wp.

The long-term aim of the project is to provide SHS to 1 million households before 2010. SELF is also collaborating with the World Bank to assess ways to integrate PV electricity into grid systems.

Source: Solar Electric Company, Washington, DC

States (funded by the US Department of Energy). The VWU is one of Vietnam's largest popular organizations, similar to the Youth Organization. Founded in 1930, the VWU has a total of 11 million members and provides programmes for women's education, childcare and local development. SELF is an NGO based in Washington, DC, which has recently established a private company, the Solar Electric Light Company (SELCO).

Initial reports from the SELF project are that the new technology has encountered some problems in providing sufficient power to meet users' needs, particularly during winter months. In addition, maintenance is also problematic, with reported failures of controllers, batteries and lamps. The ability of local technicians to deal with these problems appears to have been relatively low, requiring some experts to travel to the villages from main cities such as Ho Chi Minh City (UNOPS, 1997). The next stage of the project, in which the power and maintenance levels are both increased, may therefore solve these problems.

Other approaches to integrating economic growth with sustainable energy strategies include the use of BOT schemes for urban waste treatment in Vietnam's large cities. In 1995, one BOT project of 20 years' duration was approved with an Indian group for using urban waste in Ho Chi Minh City to generate electricity and manufacture organic fertilizer (*VIR*, 27 March–2 April 1995). This policy has allowed Vietnam to develop renewable energy resources and deal with growing urban waste in a way which transfers the infrastructure to the Vietnamese government. The contact with India is also a good example of South–South technology transfer, and was encouraged with Vietnam because of the two countries' international support for each other during the 1970s following the communist victory in Vietnam.

Horizontal transfer

A variety of ODA and pilot development schemes have attempted to increase the adoption and awareness of renewable energy projects in Vietnam. The Institute of Energy, part of the Ministry of Industry, has conducted research on new forms of renewable energy and their possible applications. This includes the Solar Lab of Ho Chi Minh City, which is part of the Institute of Physics in the Vietnam National Centre for Natural Science and Technology. Solar Lab's first PV battery charger, with a capacity of 300 Wp, was installed in 1990, and by mid 1997 had been extended to 33 villages in the Mekong Delta area in conjunction with French ODA.

The Ministry of Agriculture and Rural Development has stated that it will work with EVN to invest US$28 million in providing up to 150 microhydro stations plus minigrids in mountainous regions of Vietnam before 2000. The long-term goal is to electrify 80 per cent of rural delta households, and 60 per cent of households in central and highland areas by 2000.

The project builds on existing work by EVN in installing 11 PV systems of 12,000 Wp each to supply telecommunications systems in remote mountainous districts between 1994–5. These projects have sought to supply government agencies and their official uses, and so are on a different undertaking to the SELF project, which sells directly to villagers. The Canadian International Development Agency (CIDA) has also been active in using PV to power water pumps in Soc Son and Son Tay districts 50 km from Hanoi through ODA schemes.

Similar measures for extending microhydro systems have been attempted through the Vietnam Bank for Agriculture and Rural Development, and the Vietnam Bank for the Poor, often in collaboration with mass organizations such as the VWU, the Veteran Association and the Farmer Association (Toan, 1997:11). The existing networks that these organizations represent allow the dissemination of renewable energy technology on lines that simply providing finance alone cannot do.

Internationally, there has also been attention to Vietnam through the establishment of the Council on Renewable Energy in the Mekong Region (CORE) in 1996. CORE was created by the a variety of research institutes and scientific departments of governments in the region to establish local 'focal points' for collecting and disseminating information on decentralized renewable energy and electricity production (Rakwichian *et al*, 1996).

Summary and conclusions

Vietnam is at an early stage of both privatization and electricity supply development. Electricity generation is still dominated by state ownership and bureaucracy, and electricity supply is dominated by large hydro projects. The government is addressing future electricity needs by commissioning large fossil fuel thermal power stations using indigenous and imported coal and natural gas reserves.

Renewable energy development, outside further large hydro schemes, is therefore secondary to providing the main requirements of power generation for Vietnam's expected industrial growth over the next two decades. Furthermore, most development schemes involving renewable energy are still considered to be fundable through ODA and to be part of centrally planned government objectives for development rather than decentralized initiatives incorporating a variety of local and international actors.

Renewable energy development and technology transfer are therefore difficult to categorize in terms of vertical and horizontal integration because of the dominant influence of government planning and ODA, and the comparative lack of direct investment by international producers (although BP Solar does have a subsidiary in Hanoi). The SELF collaboration with the VWU represents one innovative attempt to introduce PV technology to the local level. Furthermore, the investment in the BOT urban waste power plant in Ho Chi Minh City by an Indian manufacturer indicates an example of South–South technology transfer of particular relevance to the expanding city's needs. However, in general, development efforts in renewable energy have concentrated more on capacity building and research from technical departments of the government, which are designed more to show the technical capability of renewable energy than its potential to be integrated with new investment and market forces.

The most important problem for renewable energy investment in Vietnam is not a lack of physical locations or market demand, but an absence of an economic or regulatory structure that can enhance its development. Negotiations between investors and government bureaucracies are in general time-consuming and costly. The experience of many joint ventures (such as the non-energy related investors Proctor and Gamble and Coca Cola) indicates that long-term investment is being discouraged because of high transaction costs and poor planning.

The isolation of Vietnam during the 1970s and 1980s has meant that it lacks many of the international organizations designed to help international investment and development. For example, the US Overseas Private Investment Corporation (OPIC) only signed an agreement to reopen its office in Vietnam after an absence of 20 years in March 1998.

Complaints by investors such as Proctor and Gamble and Coca Cola may be claimed to be an attempt to negotiate more independence and a freedom from having to transfer technology horizontally through joint ventures. This indeed may be the case, but the widespread existence of similar reports from other companies suggests that the current interface between foreign investors and government bureaucracies is resulting in impasse rather than an acceleration of both investment and technology transfer. Paradoxically, Vietnam's reputation as a location of strict command-and-control legislation is failing because so little of these regulations can be enforced.

Renewable energy investors have shown that some power development may be achieved through using existing networks of social and political organization such as within organizations like the Women's Union, or the international linkages between India and Vietnam. The small steps towards privatization and liberalization undertaken so far within Vietnam have not yet provided a major opportunity for providing widespread renewable energy investment, but have contributed to rapid social and economic change that have eroded the political ability to enact reform. At present, the best hope for building renewable energy investment in Vietnam appears to come from the concerted action of international and local development organizations to ensure that new institutions are created to enforce the adoption of renewable energy applications within rural development schemes and urban waste management.

Chapter 8

Renewable energy investment under strict bureaucracy: the case of Indonesia

Introduction

This chapter continues the analysis of renewable energy investment and technology transfer by focusing on the circumstances in category 3 of Table 5.1. Under this category, renewable energy technology is likely to be in great demand because of a lack of a centralized grid. However, foreign direct investment is made difficult by the existence of a regulatory bureaucracy that may enforce horizontal rather than vertical forms of integration and technology transfer. This strong government intervention may, none the less, make state-led renewable energy ventures or incentives more common.

Indonesia is the case study chosen for this chapter. Like the Philippines, Indonesia is a vast archipelago with many small islands making it expensive or impractical to extend a centralized grid. However, unlike the Philippines, the government has yet to introduce or encourage large-scale privatization and liberalization of the electricity supply industry that actively focuses on renewable energy. In addition, Indonesia is in the paradoxical situation of having more than 13,000 islands, but some 60 per cent of its rural population living on the two most central and grid-supplied islands of Java and Bali. The majority of rural settlements on the outer islands are dispersed and remote, thus adding to electrification costs. In 1990, the ADB estimated that just some 25 per cent of Indonesia's rural population had access to grid-supplied electricity.

This chapter starts with a brief review of the energy trends and structure of the electricity market in Indonesia. Recent attempts at privatization and liberalization are then reviewed prior to a more complete analysis of the impacts of market structure on renewable energy investment and technology transfer. The conclusion summarizes what lessons can be drawn for the book's wider aims of harnessing private investment for environmental policy and enhancing technology transfer.

Energy supply and renewable energy resources in Indonesia

Overview

Indonesia is the largest country in South-east Asia and the world's fifth largest country by population. Some 190 million people and more than a hundred ethnic groups live in this archipelago of 13,670 islands that extends approximately 5,100 km from east to west, and about 1,600 km from north to south. The islands are all tropical and feature forests, mountains and volcanoes. There are some 200 active volcanoes in Indonesia. The climate of Indonesia is hot and humid, with average temperatures ranging from 23 °C to 30 °C. There is heavy precipitation throughout the year, especially during the summer months of June and July. Typhoons can also affect coastal sites in Indonesia.

Indonesia declared independence from Dutch colonial rule in 1945, and attempts have been made since to unify the vast and varied country under a programme of rapid development. One persistent problem facing the economic development of Indonesia is the concentration of population within the two central islands of Java and Bali. The Transmigration Programme has attempted to address this concentration of population by taking rural and urban poor from Java and Bali and setting up new settlements in Indonesia's outer islands. Yet despite these controversial efforts at development, some 60 per cent of Indonesia's rural population still live in Java and Bali, and the majority of rural settlements on the outer islands such as Sumatra, Kalimantan and Irian Jaya are small and dispersed.

Indonesia is not as developed as its neighbouring countries of Thailand and Malaysia, and indeed many of the outer islands of Indonesia such as Irian Jaya have very low levels of urbanization, industrialization or modern education. During 1980–94, the Indonesian economy grew at an average of 5.5 per cent, helped partly by a stringent monetary policy that encouraged inward investment. However, since 1997

Indonesia has suffered a severe economic reversal as a result of poor financial management and a collapse in the currency, the Rupiah (Rp). There has also been political instability following the fall of President Suharto in 1998 who had ruled almost dictatorially since 1965. Suharto's form of absolute government contributed to an old fashioned and bureaucratic administrative system, plus fears from investors concerning possible further political disturbances.

Indonesia remains a largely agricultural country. Agriculture accounts for a quarter of GDP, but employs more than half of the population. Indonesia is also South-east Asia's largest exporter of coal, natural gas and oil, although its income from these exports has been affected badly by the decline in oil prices since the mid 1980s. In the fiscal year 1983–4, oil and gas sales contributed 66 per cent to government revenue. In 1996–7 the budgeted amount was just 18.1 per cent. ODA from other countries has remained a significant, if variable, source of finance. In 1983–4 ODA was equivalent to 26.9 per cent of domestic revenues; in 1990 it reached 32.8 per cent and in 1996–7 it was budgeted to be equal to 15.9 per cent (Ministry of Finance, in Symon, 1997:23).

Energy trends

Indonesia is comparatively well endowed with oil, natural and coal resources. Proven oil reserves amount to 10 billion barrels, with probable reserves ranging from 10 to 40 billion barrels. Current production is 1.9 million barrels a day. Natural gas reserves total 2,100 billion cubic metres, with an estimated extra potential supply of some 653 billion cubic metres. Over 90 per cent of natural gas is non-associated with crude oil, and so production can take place independently of oil. Coal resources are estimated at 26 billion tonnes, consisting mostly of sub-bituminous and bituminous coal in west and south Sumatra and in east and south Kalimantan. Government policy has attempted to maximize the use of indigenous energy resources, with the result that coal production rose from less than 1 million tonnes in 1983 to 7 million tonnes in 1989 (Green, 1997:37).

Table 8.1 indicates projected energy demand for Indonesia 1995–2010. The greatest increase is set to come from coal, which is projected to grow by six times by 2010. After this, natural gas consumption is expected to increase by three times. Oil, geothermal and hydroelectric power to increase by two-and-a-half times. Indonesia also has a nuclear power programme, which is expected to start producing electricity by 2010.

Table 8.1: Indonesia: primary energy demand by source, 1995–2010 (mboe)

	1995	2000	2005	2010
Oil	306,782	421,600	544,547	727,153
Natural gas	104,755	143,961	227,720	341,317
Coal	52,377	71,981	148,513	311,637
Hydro	29,930	41,132	59,405	81,619
Geothermal	4,988	6,855	9,901	10,388
Nuclear	–	–	–	11,872
TOTAL	498,833	685,529	990,086	1,483,986

Source: Indonesia Ministry of Mines and Energy (in Symon, 1997:30)

Renewable energy resources

Indonesia has great potential for both hydropower and geothermal power. The total hydropower potential is estimated to be 75,000 MW, of which 34,000 MW is exploitable. However, only 5 per cent of these resources are located in Java, which currently accounts for 70 per cent of electricity demand. Irian Jaya and Kalimantan have 59 per cent of resources but only just 4 per cent of total consumption. In 1990, total installed capacity of hydropower plants was 3,000 MW. However, as in Thailand, there has also been great local opposition to large dam projects. For example, the proposed Jatigede Dam, 140 km south-east of Jakarta is intended to be a multipurpose $600 million project contributing to power generation, irrigation and urban water supply. However, local activists have demonstrated against the proposed relocation of 6,500 households and flooding of 4,000 ha, potentially forcing the World Bank to pull out of the project (Symon, 1997:122).

Geothermal potential is estimated to be 16,000 MW, but resources are much closer to demand centres than with hydropower. There are approximately at least 5,500 MW in Java, 1,400 MW in Sulawesi, 1,100 MW in Sumatra and 2,300 MW on other islands. There are more than a hundred volcanic cones in Java, of which 15 are active and 7 under observation (Green, 1997:38).

There is also great potential for smaller renewable energy technology. Table 8.2 lists the potential of various hydropower projects in Indonesia, including small hydropower. It is generally recognized that mini- or microhydropower devices have great potential in the rural settlements of Indonesia, especially where these can be integrated with local irrigation systems or rice mills. Similarly, biomass is already extremely important in rural energy systems, believed to be around 40 per cent of existing fuel use. An estimate in 1992 placed the known reserves of solid biomass that can be used on a sustainable basis to be about 246.6 billion tonnes per year (Green, 1997:52).

Table 8.2: Summary of potential hydropower potential in Indonesia

Island	No. of sites	Potential (MW)	Energy (GWh year^{-1})
Java	149	4,531	18,042
Sumatra	174	15,804	84,110
Kalimantan	177	21,611	107,202
Sulawesi	116	10,203	52,952
Irian Jaya	210	22,371	133,579
Bali and NU.S. a Tenngara	136	674	3,287
Maluku	53	430	2,292
TOTAL	1,015	75,624	401,464

Source: Water Power and Dam Construction, Current status of hydro development in Indonesia, Nos 45/4, April 1994 (in Green, 1997:50)

Average daily solar radiation in Indonesia is approximately 4 kWh m^2, and so PV and solar energy devices have clear potential. The potential for wind energy is less clear. Some early attempts to build wind power suffered from generally low wind speeds of only between 3 and 5 m s^{-1}, plus corrosion of equipment because of salt water. However, building wind energy has suffered from an inadequate mapping of wind

Renewable energy investment under strict bureaucracy 147

resources, and now several projects are adopting wind energy in a variety of locations.

Structure and liberalization of the electricity supply industry
Electricity market and institutions

The Indonesian electricity supply industry is characterized by centralized grid systems in the islands of Java and Bali, and a diversified supply of electricity to the outer islands. At the end of 1996, the generating capacity of the national utility, Perusahaan Listrik Negara (PLN) was 15,000 MW. This represents impressive growth from just 500 MW in 1970. At the end of 1995, the installed capacity was 46 per cent oil and diesel, 19 per cent natural gas, 18 per cent coal, 15 per cent hydroelectric power and 2 per cent geothermal. In addition to this, there are a vast number of private 'captive' generators owned by companies or households in order to guarantee their own energy supply. These private generators amount to the equivalent of another 40 per cent of the national PLN capacity, and generally use diesel as a fuel source (Symon, 1997:31).

Table 8.3 shows planned capacity additions according to generation technology and private-sector involvement. This shows large capacity additions planned for steam, gas turbine and combined cycle generation, as well as in hydroelectric power. The aim is to almost triple electricity generation capacity from 13,176 MW in 1994 to 34,413 in 2004.

However, it is still unclear how much of this electricity will reach the outer islands that are currently unconnected to the central grid. Table 8.4 indicates the status of electrification in Indonesia. This indicates that some 53 per cent of rural households were without official grid supply from PLN in 1995, and that the majority of rural households depend on kerosene as the main fuel for lighting. Most households classified as urban are connected to PLN grid electricity. However, figures for rural electrification vary greatly between different provinces. In Irian Jaya in 1995, for example, it was estimated that 82 per cent of the population had no access to electricity. This compared with 81 per cent for East Timor; 73 per cent for central Kalimantan; 76 per cent for south-east Sulawesi; 59 per cent for west Sumatra; but 36 per cent for west Java, and 18 per cent for Bali (Symon, 1997:49).

Table 8.3: Indonesia: projected electricity generation technology and private-sector involvement, 1994–2004 (MW)

Plant type	Total capacity 1994	Capacity additions 1994–99		Capacity additions 1999–2004		Total capacity 2004		
	All (PLN)	PLN	IPP	PLN	IPP	PLN	IPP	All
Hydro	2,215	821	–	3,392	–	6,428	–	6,428
Steam	4,341	2,595	1,630	1,375	4,520	8,311	6,150	14,461
Combined cycle	2,817	3,139	1,098	66	66	6,022	1,164	7,186
Geothermal	250	353	135	58	895	661	1,030	1,691
Gas turbine	1,413	1,261	–	160	–	2,834	–	2,834
Diesel	2,140	420	–	10	120	2,570	120	2,690
Microhydro	–	123	–	–	–	123	–	123
TOTAL	13,176	8,712	2,863	5,061	5,061	26,949	8,464	34,413

Note: dates refer to fiscal year April 1 to March 31

Source: PLN (in Symon, 1997:41)

Table 8.4: Indonesia: electricity supply to urban and rural households, 1995

	Total population (millions)	Households with PLN electricity (%)	Households with non-PLN electricity (%)	Kerosene and other lamps as source of light (%)
Urban	70.5	92.14	1.25	6.61
Rural	124.8	47.45	5.09	47.46
National	195.3	62.95	3.75	33.30

Source: Indonesia Central Bureau of Statistics (in Symon, 1997:48)

Indonesia has consequently paid recent attention to improving transmission and distribution. Between the years 1988 and 1994, transmission and distribution networks increased at the rate of 6 to 7 per cent annually, but generation still outstripped this at an average 11 per cent per year. As a result, some power plants are running below capacity because it is not possible to transmit or distribute the power. Indeed, the World Bank has estimated that 15 per cent more electricity sales

Renewable energy investment under strict bureaucracy 149

could be made in Indonesia at present through better transmission and distribution. In addition to this, PLN has also estimated that some 20 per cent of potential sales fail to come about because of an inability to deliver, theft from unauthorised users, and the trend towards foreign investors building their own captive plants. However, system efficiency is apparently improving. In the 1980s, some 20 per cent of electricity on the Java–Bali grid was lost through transmission. In the mid 1990s this had fallen to 12 per cent (Symon, 1997:47). Table 8.5 details planned expansions to transmission and distribution lines in Indonesia, 1994–2004.

Table 8.5: Indonesia: planned expansions to transmission and distribution lines 1994–2004

Facilities	Total 1994	Additions 1994–1999	Additions 1999–2004	Total 2004
Transmission lines (km)	19,869	10,548	15,390	45,807
Transmission substations (MVA)	23,936	30,406	26,080	80,422
Distribution lines, medium voltage (km)	118,315	133,319	144,295	395,929
Distribution substations (MVA)	17,899	21,824	23,386	63,109
Distribution lines, low voltage (km)	162,442	196,741	215,290	574,473

Source: PLN (in Symon, 1997:48)

The management of the Indonesian electricity supply industry is undertaken by the Ministry of Mines and Energy (MME), and the state electric utility PLN. Government economic planning is conducted in a controlled manner involving five-year plans (called Repelitas) which started in 1969. In addition, there are also 25-year long-term strategies of which there have been two so far, from 1969 to 1994, and 1994 to 2019.

Energy policy is enacted by PLN, but also in conjunction with the state oil and gas operator, Pertamina. Pertamina has both a practical role as an oil and gas extractor and distributor, plus also a regulatory role through determining the price at which natural gas is sold to different industries by the private sector. Furthermore, other relevant companies include the state coal mining enterprise, PT Tambang Batubara Bukit Asam (PTBA); the domestic gas transmission and distribution company, PT Perusahaan Gas Negara; and the energy efficiency advisory company, PT Konservasi Energi Abadi ('Koneba'). The administration of hydroelectric projects is undertaken by the Ministry of Public Works on account of the multiple uses for dams including water supply and irrigation.

The MME is the main regulator of energy policy and the PLN, and enacts decisions through the Directorate General for Electricity and Energy Development (DGEED). The directorate sets retail and wholesale energy prices, and advises PLN on a variety of technical and financial issues. In addition to this, the Ministry of Finance regulates PLN concerning the provision of large loans of more than Rp 3 billion (or US$1.5 million). All major investments and decisions involving power generation have also to be considered by three coordinating agencies. These are the Tim Koordinasi Pengelolaan Pinjaman Komersial Luar Negeri (PKLN), which provides advice on managing offshore commercial loans; the Baden Koordinasi Penanaman Modal (BKPM), to assess capital investment in power and mining (Pertamina assesses specific oil and gas projects); and Baden Engendalilan Dampak Lingkungan ('Bapeda'), which is the environmental standards and impacts agency.

In the late 1970s, a specialist agency for providing research and assistance with alternative and renewable energy sources was established: Baden Pengkajian dan Penerapan Teknologi (BPPT). The activities of this are discussed more in section 8.4.

Privatization and liberalization

The Indonesian electricity supply industry has one of the largest potential markets in South-east Asia, yet its generation and regulation structure has remained largely centralized and bureaucratic. Indeed, until Law No.15 on electric energy in 1985, the organizations administering the Indonesian electricity supply had remained broadly unchanged since the end of colonial rule in the 1940s.

Law No.15 was the first general piece of legislation to set up the possibility for new regulation and supply of electricity within Indonesia. This was followed in 1989 and 1990 by Government Regulations Nos 10 and 17, which confirmed the central role of PLN, but also allowed generation of electricity by licensed private sector companies and cooperatives for public use. Further important reforms included Presidential Decree No. 37 (1992) on the private sector undertaking for electric power supply, and the Minister of Mines and Energy Regulation No. 02.P/03M.PE/1993, which further enabled private enterprises and cooperatives to generate and distribute electricity.

Since the passing of these reforms in the early 1990s, the participation of private investors in electricity generation has increased, although to date there has only been one private company who has invested in transmission (in southern Sulawesi). This investment, however, was in potential conflict with the Indonesian Foreign Investment Law (Law No.1 of 1967), which stated that foreign investment was allowed in energy production or distribution. It was therefore necessary to undergo the lengthy process of amending this law. In June 1994, Government Regulation No. 20 on share ownership in foreign invested companies was passed, plus the Decree of the Minister for Investment No. 15/SK/1994 on share ownership amended the regulations of 1967, and therefore made it legitimate for foreign investors to take ownership of electricity production and supply.

The first PPA for a BOT power project was signed in February 1994 for a coal-fired plant at Paiton in Java, east of Surabaya. The project was a joint venture between PLN and PT Paiton Energy Co., of which partners included Mission Energy and General Electric (of the United States); Mitsui (of Japan); and PT Hitam Perkasa (Indonesia). By the end of 1996, 13 foreign companies had signed memoranda of understanding to undertake IPP projects in Indonesia.

There have also been attempts to liberalize the structure of PLN as well as encourage private and foreign participation in power production. Government Regulation No. 23 in August 1994 transformed the PLN from a public sector enterprise to a limited liability company PT PLN (or 'Persero'). After this ruling, PLN was no longer able to assume funds from the Ministry of Finance for capital development or the support of current expenditure. However, PLN could establish joint ventures with the private sector and also sell assets and raise funding nationally and internationally. Again, this legislation came some time after actions legitimized by it had already taken place. The PLN had already sought funding by issuing bonds domestically in 1992 to the value of Rp 300 billion (US$150 million). A year later, Rp 600 billion ($320 million) was raised, and then Rp 700 billion (US$375 million) in 1995 and Rp 1 trillion (US$600 million) in 1996.

In 1995, PLN established two new operating subsidiaries for the east and west sections of the Java–Bali grid. A third subsidiary for geothermal generation is also being considered. The two subsidiaries of Java–Bali also seek to sell shares direct to the stock market by the late 1990s. Following this new market orientation of PLN, the responsibility for negotiating PPAs with private sector bodies has been passed from DGEED to PLN. Furthermore, in October 1994, the government revised tariffs every three months in order to reflect costs and competition from other suppliers (Marsudi, 1997; Symon, 1997:33–7).

However, despite this progress towards efficiency and privatization, critics have suggested that the PLN has been paying too much for IPP-generated power. Furthermore, it is also generally recognized that more attention needs to paid to encouraging foreign investment in transmission and distribution. Indeed, the PLN runs the risk of having to reduce its own low-cost power generation in order to honour its PPAs with IPPs at times when the transmission system is fully employed. Consequently, PLN is facilitating privatization of electricity generation at a risk to its own operation, and paying for IPP electricity that cannot all be used. Furthermore, the oil and gas enterprise, Pertamina, has also been criticized for maintaining control over the price of natural gas production by the private sector because it adds to private investors' risk, and also indirectly influences electricity pricing (Symon, 1997:21, 41).

Renewable energy and technology transfer

Vertical transfer

Vertical technology transfer, or point-to-point technology investment are difficult to quantify in Indonesia because much renewable energy investment is linked to ODA. Much ODA may attempt to introduce renewable energy technologies to new regions, but this kind of point-to-point relocation is also associated with deliberate government policies for rural development rather than the commercial operations of companies. Furthermore, ODA-related investment is also linked to the long-term building of capacity for adopting technology in the form better described as horizontal integration.

Nevertheless, there has been some activity in introducing new renewable energy technologies through direct investment. Recent changes in the rules for private ownership have allowed foreign companies to undertake BOT projects for geothermal electricity production. The first commercial geothermal generation started in 1982 on an old site at Kamojang in west Java with partial investment by Mitsubishi Heavy Industries (Japan) plus ODA from New Zealand and assistance from the World Bank.

Geothermal electricity production is hampered commercially by the difficulties in establishing fair prices when there is no international spot market. In addition, Presidential Decrees Nos 22 (1981) and 45 (1991) established Pertamina as the sole authority to liase between potential investors and PLN for setting prices and contractual obligations. These controls increase the amount of bureaucracy involved in investment, and reduce the autonomy of investors (Symon, 1997:114).

Similarly, foreign investment in hydropower is still relatively limited because the MME and PLN share joint responsibility for the operation of dams with the Ministry of Public Works. This complicated arrangement is because of the variety of uses to which dams may be put. Foreign involvement in ocean and wave technology, however, has been encouraged by BPPT. A US$7 million 1.1 MW demonstration project has been constructed on the southern coast of Java at Baron Beach near Yogyakarta (Symon, 1997:129). There are also possible opportunities for foreign direct involvement in smaller renewable energy technologies.

However, small-scale renewable energy development has been encouraged by the government's specific PPA for small power producers, the Pembangkit Skala Kecil Swasta dan Korporasi (PSKSK). The aim of the PSKSK is to encourage developers to use fuels besides oil. Under the PSKSK, small power producers are allowed to generate and sell electricity to PLN in amounts of up to 30 MW to the Java–Bali grid, and up to 15 MW for other systems. The tariff is ranked by type of energy input, with wind, solar and minihydro given a higher standing than oil, coal and gas. Biomass fuel, including vegetable and animal waste is of secondary importance.

Box 8.1 provides details of one international investor's approach to building biomass generating plants in Indonesia. The investor, Bronzeoak of the United States, undertook a broad evaluation of the Indonesian market for biomass with particular focus on the palm oil sector. The World Bank estimated in 1994 that a surplus of up to 247 MW could be produced countrywide by palm oil waste (Walden, 1997:1). It has been estimated that a 10,000 hectare oil palm plantation produces enough waste to support a 7.4 MW power plant at a relatively low plant heat rate of 21,100 kJ per kWh. However, the initial investment has also been hampered by difficulties in negotiating a PPA with PLN.

Box 8.1: Bronzeoak's biomass investment in Indonesia

Initial approaches to investment were prepared in consultation with the US Export Council for Renewable Energy, and market and resources reports from Winrock International, a US-based non-profit making development organization. Local market surveys were carried out in conjunction with the Environmental Business Support Foundation of Jakarta.

The plant was designed by the McBurney Corporation of the United States, for a 36,450 kg hr^{-1}, 4.58 Mpa, 400 °C boiler with a steam turbine for extraction and condensation. This plant could produce surplus electricity of 41.8 million kWh, with empty fruit bunches, fibre and shells used as fuel. The maximum generation capacities range from 7.8–12.3 MW, although average rates are between 4.8–9.2 MW. The plant can also operate on alternative fuels, such as palm fronds, waste palm trunks and wastes from other plantations within the distance of economic transportation.

The technical capability of the plant is however, not matched by negotiations over funding. Bronzeoak aims to assure the project financially by negotiating a PPA with PLN, which will enable Bronzeoak to raise debt finance of about 70 per cent of the cost of the project based on the collateral of the project itself. However, the PLN had (by August 1997) not agreed to this, although negotiations were still under way. The situation was helped by the decision by the World Bank to approve a loan of more than US$66 million for technical support and debt financing of biomass projects. The World Bank has therefore sought to influence PLN to accept PPAs at the levels asked for by investors.

Source: Walden, 1997

A more grand-scale opportunity for vertical technology transfer is the so-called 'one million homes photovoltaic rural electrification project in Indonesia' (sometimes also called the 'Fifty Megawatt Peak Photovoltaic Rural Electrification Programme') launched in June 1997 by the Indonesian government. Since 1988, more than 10,000 Solar Home Systems (SHSs) for basic lighting have been installed in rural locations as an interim measure before grid extension. BPPT aims to provide a further 36,400 SHSs, or one million rural households with SHSs by 2007 (or about 10 per cent of non-electrified households) at an estimated cost of $450 million. This, however, is not totally controlled by private investment and is generously supported by the World Bank and ODA assistance (see Box 8.2).

Horizontal transfer

The Indonesian government has undertaken a number of important steps to increase the adoption of renewable energy technologies. On many occasions, these have included encouraging actions by foreign direct investors. However, in general these actions have been coordinated into wider-scale projects to integrate these new technologies into rural economies and development programmes. As a result, it may be more constructive to view this combination of private-sector investment and government organization as a form of horizontal integration. This has been particularly apparent in the development of small renewable energy technologies for rural electricity supply.

PLN has overall authority for managing rural electrification and small-scale renewable energy development. However, much of the driving force behind recent initiatives has been BPPT, which has favoured a strong interventionist stance in rural energy backed up by research and development. The specific focus of BPPT has allowed rural development programmes to overcome some historic problems such as the diversified nature of many rural settlements in Indonesia's outer islands, and the high installation and lack of flexible billing historically associated with PLN.

> **Box 8.2: The one million homes PV rural electrification scheme in Indonesia**
>
> The scheme was launched in June 1997 with a long-term goal of providing electricity to one million households, or about 10 per cent of non-electrified households by 2007, at an estimated cost of $450 million. The technology chosen is Solar Heating Systems (SHSs) with an aim to provide 50 Wp per household needs such as lighting, and radio or television operation. The SHS technology is constructed in Indonesia.
>
> The project has not been implemented a purely commercial basis, and support comes from the World Bank, GEF, the Australian international aid agency, AusAid, France and the German state of Bavaria. The initial focus has been on the islands of eastern Indonesia including Sulawesi, Maluku, west and east Nusa Tenggara, eastern Kalimantan and Irian Jaya.
>
> Financing has been organized to offer a gradual transition of ownership of the SHS from the government to the households. Furthermore, the project is divided into three schemes to indicate which households are furthest from the central electricity grid, and who may have to wait the longest to be attached to this grid. In general, those households that are furthest from the grid have the most attractive payment terms in order to encourage adoption of the SHS.
>
> The scheme is held to be ambitious in both technical and developmental terms. However, three main obstacles to success have been identified. First, the SHS technology needs a high initial investment and is being disseminated to locations where people have low spending power. Secondly, the quality of communication, and financial and technical management in the selected areas are also very poor. These problems may mean the scheme will encounter problems of maintaining both technical performance and regular payments. Thirdly, the main equipment for SHS such as inverters and battery charge regulators will have to come from Jakarta, thus increasing costs and delays. As a result of these three problems, it is clear that the one million homes electrification scheme will only succeed with efforts to combine vertical technology transfer of PV with horizontal efforts at capacity building and local maintenance skills. BPPT has already proposed and developed a network of management information systems. But the ultimate success of this far reaching scheme is not yet guaranteed (but see Box 8.4).
>
> *Source: Djojodihardjo et al (1997)*

Hybrid systems, combining renewable energy technology with diesel, have also been encouraged. BPPT has worked with Integrated Power Corporation (IPC), a subsidiary of Westinghouse Electric of the United States, in combining PV and wind turbines with diesel. An important aspect of ensuring success has been household pre-payment systems using meters in which villagers may use disposable magnetic strip cards (like telephone cards) purchased from local shops as a way to trigger the device (Symon, 1997:129). Hybrid systems may also contribute to local mini-grids of between 1 and 12 MW capacity.

The state gas company, PGN, has also worked collaboratively with the German company Gotz Gmbh on a feasibility study for using urban waste in electricity generation. The aim is to ferment organic waste in order to produce methane and carbon dioxide at Bantar Gebong in Bekasi west of Jakarta. The estimated cost of this is between US$20 and US$40 million. However, this project represents an example of horizontal integration through requiring Gortz Gmbh to form a consortium with local private companies in order to pass on experience (Symon, 1997:129).

Some foreign investors have also employed small hydro systems. Most small hydro devices are less than 50 MW in capacity, partly due to the need to integrate power generation with local irrigation networks in many circumstances. The government aims to increase small hydro capacity from 2,196 MW in 1999 to 6,803 MW in 2004. As part of this existing capacity, some 70 plants of minihydro (of between 200 kV and 5 MW) with a total capacity of 55 MW have been recorded, although some date from the time of Dutch colonialism. However, larger joint ventures often linked to mining activities are increasing. For example, in northern Sumatra, a Japanese–Indonesian joint venture has created two power plants of 268 MW and 317 MW on the Asahan River next to the PT Asahan Aluminium smelting works. In southern Sulawesi, a similar Canadian–Indonesian joint venture has built a 165 MW plant on the Larona River to service a local nickel plant and sell surpluses to the public grid (Symon, 1997:123).

Similarly, wind power generation has been encouraged in sites in eastern Indonesia where wind speeds above 4 m s^{-1} make wind turbines competitive with diesel. The government National Aeronautics and Aviation Agency has facilitated two pilot projects in villages on the northern coast of central Java. In the villages of Bulak and Kalianyar there are some 1,000 households with more than 30 wind turbines, supporting local industries such as furniture manufacturing.

However, the incorporation of local users in renewable energy use depends partly on the ability to ensure that the technology is adopted on terms that are practically and economically feasible. Much research has been undertaken on which financing and implementation measures are necessary to ensure maximum success of investment in renewable energy. Indonesia's one AIJ project is an attempt to do this. The Australian International Centre for the Application of Solar Energy (CASE) has undertaken a scheme in order to build capacity for adopting renewable energy technology under AIJ in Indonesia. The aim of this is to work in conjunction with BPPT to deliver appropriate technology training to local users of new technology, and then demonstrate its use via a pilot project (*International Solar News*, August 1997, p8).

The project can take lessons from other practitioners of renewable energy technology. The US-based development organization Winrock International has successfully implemented wind turbines in the islands of Eastern Indonesia through a process of technical and commercial extension. Wind turbine technology has been used for water pumping for many years. However, sophisticated machinery breaks easily, and the equipment can be broken up and parts stolen during the time it takes to be repaired. Winrock have attempted to overcome this by helping establish a local, or distributed, utility within villages to manage the project in technical and commercial terms. The major problem is in introducing the concepts of paying for water and power used per household. However, the cooperation between the utility, Winrock, and a local NGO ensures that the tasks are achieved (see Box 8.3).

This kind of local implementation has also been demonstrated in SHS technology. SHS has been discussed above in the context of vertical integration, but there has also been work on implementing this new technology in association with local villagers and finance sources. Box 8.4 presents a case study of the Sukatani village in west Java in which some success has been achieved in establishing both renewable energy technology and financial management.

Winrock and USAID also helped set up the Renewable Energy Network of Indonesia (RENI) in May 1996 in collaboration with an Indonesian NGO, Yayasan Bina Usaha Lingkungan. The aim of the network is to stimulate the development and implementation of renewable energy in Indonesia by private-sector investors and joint ventures through project identification, cost sharing and training in new technologies (see RENI, 1997). This information has helped

> **Box 8.3: Winrock International management of wind turbines in Eastern Indonesia**
>
> Winrock International is a non-profit making development organization. In 1995 it installed ten wind turbines in the islands of Eastern Indonesia for small-scale rural enterprises. Of the ten turbines, six were for water pumping, and four for power generation. All were constructed by Bergey Windpower of the United States, and were between 1.0 and 1.5 kW power supply.
>
> The project did not simply attempt to implement the technology, but also to set up a monitoring and governance mechanism to ensure that the energy supply was maintained in years to come. One common problem with new technology is the rapid breakdown and then theft of parts of equipment. In order to overcome this, Winrock began the establishment of a local electricity utility within the villages. This was a difficult task as it involved measuring and then requiring payment for electricity and water used, both of which were new to the villages. The tariff charged also included amounts for maintenance and the supply of parts directly from the manufacturer.
>
> The monitoring of the utility was carried out through a tripartite arrangement between Winrock, the utility and a local NGO. Each were assigned duties concerning the formulation of financial reports, training in management and maintenance, and collection of payments. The aim of the tripartite relationship is to ensure that local participation is assured in the establishment and maintenance of the project, but also to regulate the new utility in its financial operation and service to users. Profits are shared, disproportionately, between the three parties.
>
> The project forms part of a programme not just to introduce new technical equipment to remote villages in Indonesia, but also to extend financial management to these locations. Water pumping and individual power usage is charged to individual users. Also, many villagers use the electricity for further money-making ventures such as selling ice drinks and popsicles. Winrock International also seek to reduce their costs by changing suppliers of wind turbines from one manufacturer to another.
>
> *Source: Winrock International, presentation to the Asia Pacific Initiative for Renewable Energy and Energy Efficiency Conference, 14–16 October 1997, Jakarta*

facilitate the growth of private-sector interest in renewable energy in Indonesia.

Summary and conclusions

Indonesia has great potential for renewable energy development on account of its large number of islands and the remote and disparate nature of rural settlements on these islands. However, the practical ability for investment to be channelled this way is limited because of

> **Box 8.4: Local implementation of solar heating systems in Sukatani, Indonesia**
>
> The first installation of SHS in Indonesia resulted from collaboration between a Dutch PV company (R&S), BPPT and the Indonesian Ministry for Cooperatives. Eighty SHS were installed in the village of Sukatani in west Java as a pilot project to demonstrate the feasibility of PV technology for rural electrification.
>
> R&S installed the SHS with the help of a local cooperative, which also provided a credit service, fee collection and simple maintenance. The equipment was PV modules of 40 Wp, a charge controller, a 12-volt battery, plus lighting and cables. Initial investment was provided by R&S, the Netherlands Ministry of Foreign Affairs, and BPPT. Villagers had to pay a down payment followed by monthly contributions.
>
> The Indonesian president, Suharto, built on this scheme by initiating the Presidential Assistance Project (BANPRES), which allowed interest-free credit for 3,000 SHSs. After ten years of monthly payments, the SHS became the property of the villagers, and local maintenance of technology was assured by keeping a small percentage of payments for contributions to a local cooperative that undertook all repair work.
>
> Since the establishment of this system at Sukatani, a similar system has been installed at the village of Lebak. By 1993, it was estimated that the average investment for SHSs had fallen below $400, for which villagers could pay $8 a month after an initial down payment of 35 per cent. One important source of core funding for these projects by donors has been the revolving fund – or a fund where running costs are met with the continuous return of loan capital with interest.
>
> Source: Gregory et al (1997:112–113)

the highly bureaucratic nature of government, and political uncertainty. Investor confidence in government energy supply is low, forcing many to adopt their own stand-alone power systems. Furthermore, the development of electricity in the remote islands is affected by the long-standing problem that most of the rural population is located on the central two islands which are well served by grids. Rural electrification in Indonesia therefore is not simply a matter of establishing decentralized renewable energy systems in the outer islands, and these may have to come second to extending grids to rural populations on the central islands.

Renewable energy investment is still primarily linked to ODA and official government schemes. There is little direct foreign investment for new technologies. Indeed, the large-scale 'one million homes' PV electrification project is using Indonesian-manufactured technology as a first choice. In addition, some foreign investment projects – such as Bronzeoak's involvement in biomass-fuelled generation – have been hampered by government bureaucracy and the need to negotiate at length with different agencies who are unwilling to agree to the investors' required commercial terms.

Some success has been achieved, however, in the horizontal integration of renewable energy technologies. The Sukatani and Winrock-sponsored wind projects in Eastern Indonesia have provided institutional support for both technical and commercial management of wind turbines and PV. These projects have demonstrated that the ability to transfer technology depends not just on ensuring that the practical maintenance of technology is achieved, but also on the transfer of commercial opportunities and management of the project finances. In effect, this is the creation of new institutional structures for technology transfer that bypass the state, and allow multinational investors the opportunity to sell products locally without needing to spend time and money in developing their own skills in market research and implementation. The actions of Winrock in starting the process of local governance of investment – through the tripartite cooperation of the utility, local NGO and itself – and in seeking competition between different international suppliers, suggest that bypassing the Indonesian state bureaucracy does not necessarily mean empowering investors over end users.

In addition, there are still important opportunities for mini and microhydro technology. These power sources may be used locally in villages, or in conjunction with mining and factory operations. Some success has been achieved in selling excess power from cogeneration industrial plants to local grids. The PSKSK legislation has been partly responsible for this on the same basis as the SPP laws in Thailand, and this suggests that the incentives for allowing companies with stand-alone power units to sell to the grid could be increased. The BPPT agency may be the institutional form in government to take responsibility for this.

The privatization and liberalization process in Indonesia, therefore, has not gone far enough in harnessing large-scale international investment in renewable energy in Indonesia, and procedures remain bureaucratic. However, it is possible to override these constraints in the cases of small decentralized projects undertaken through specialist intermediary organizations such as Winrock, or by encouraging private factories to sell surplus hydro or biomass power from their own generation units to local grids.

Chapter 9

Off-grid renewable energy under active investment: the Philippines

Introduction

This chapter concludes the case studies in Part II by looking in detail at category 4 of the market conditions described in Table 5.1. Under these conditions, renewable energy investment is encouraged by the market demand from large rural areas not currently served by a grid electricity supply from fossil fuel sources, and by a government policy that encourages foreign investment and private ownership of electricity generation.

The Philippines is a relevant case study for this chapter because there are many small islands and remote rural areas within the archipelago which are suitable for electrification using renewable energy technology. Furthermore, the government has recently adopted aggressive electricity privatization and electrification incentives. However, it is acknowledged that there are other locations within South-east Asia which may also suit this description, and that much of the larger islands within the Philippines such as Luzon and Mindanao are also heavily grid dominated and use fossil fuel sources.

The chapter starts with a basic description of the market and energy trends of the Philippines and then goes on to assess in more detail the structure and current restructuring of the electricity market as a prelude to analysing the growth of renewable energy. The impacts of these changes on foreign investment and technology transfer of renewable energy are then discussed. The chapter concludes by assessing what institutional factors have assisted or hampered the growth of renewable energy. The conclusions for the region, and incorporating businesses and investment into environmental policy are then discussed in more length in the final three chapters comprising Part III of the book.

Energy supply and renewable energy resources in the Philippines

Overview

The Philippines is an archipelago of more than 7,100 islands. The country is divided into three main island groups. The first of these is Luzon, also comprising the large island of Palawan, is the largest with some 35 per cent of total land area and is also the location of the capital city of Manila. The second island group is Mindanano, which also includes the islands of Sulu and Tawi-Tawi. The third island group is the Visayas, and comprises Cebu, Bohol, Panay, Samar, Negros and Leyte. The topography of the Philippines includes volcanoes and mountains, as well as coastal plains and a complex coastline for most islands. The climate is typically tropical, with an average annual and largely unvarying temperature of 27 °C, and winds affected by dry and wet monsoons and occasional typhoons. The dry season is November to May, while most rain falls during June to September.

The Philippines is one of the less developed countries of South-east Asia. A variety of factors have combined to slow economic development. These include the country's relatively small endowment of natural resources in comparison with neighbouring countries; the diversified distribution of islands and peoples; and a history of colonial and oppressive political regimes such as under Ferdinand Marcos during the 1970s and 1980s which created profits for rulers rather than provided equitable development for the rest of the population. Real GDP grew at some 3.3 per cent between 1985 and 1994, yet in 1991 44.5 per cent of the 70 million population lived below the poverty line. Manufacturing has contributed most to GDP, at some 25 per cent in 1994, followed by agriculture at 22 per cent. Most exports are electronic equipment and textiles. However, more than 40 per cent of Philippine people work in agriculture, many of them as small or tenant farmers (IEA, 1997:204; Green, 1997:83).

Energy trends

The Philippines is comparatively poorly endowed with natural energy resources within South-east Asia. Table 9.1 presents the total energy supply projection for the Philippines from 1996 to 2010. The immediate energy supply problems faced by the Philippines are a dependency on imported oil and coal, and lack of infrastructure to develop its own resources, particularly natural gas, geothermal activity and hydroelectric power. Current coal reserves are estimated at 1,600 million tonnes, of which just 268 million tons are considered recoverable. In 1990, the country produced only 1.2 million tons of its total requirement for 2.2 million tonnes, the rest being supplied by imports (Green, 1997:83).

Table 9.1: The Philippines: energy supply projection 1996–2010

Fuel	Unit	1996	2000	2005	2010
Oil	Production (m bbls)	0.48	8.10	16.56	2.22
Natural gas	Production (bn cu ft)	0.92	0.73	228.13	355.88
Geothermal	Installed capacity (MW)	1,414	2,014	2,134	2,859
	Cumulative no. of wells	613	685	916	1,049
Coal	Proven reserves (cumulative mt)	427	610	701	917
	Production ('000 t)	2,866	4,216	7,706	10,113
Hydro	Installed capacity (MW)	2,333	2,572	4,071	4,570
	Gross generation (GWh)	5,699	6,443	11,861	13,770

Source: The Philippines Department of Energy (in Lefevre et al, 1997a:36)

Energy demand is predicted to rise 7.1 per cent between 1991 and 2000. In 2000 total demand is expected to be 219.8 million barrels of fuel oil equivalent, or double the 1991 level. Geothermal energy is expected to supply 12.9 per cent and non-conventional sources such as biomass will contribute 11.5 per cent. Imported oil, however, will remain the main supply for demand – contributing an estimated 50.8 per cent by 2000, although just 16 per cent for power production. Nuclear power is a possible source of supply, although the opening of the Philippines' only plant at Bataan was deferred after the fall of President Marcos following enquiries into quality control and corruption

Table 9.2: The Philippines: projected energy consumption 1992–2000

	1992	1995	2000
Indigenous energy (%)	35.1	37.5	38.0
Oil and gas	2.1	2.2	0.7
Coal	3.9	6.0	7.0
Hydropower	8.4	7.1	5.8
Geothermal	7.8	9.9	12.9
Non-conventional	13.0	12.3	11.5
Imported energy (%)	64.9	62.5	62.0
Oil	59.9	54.8	50.8
Coal	5.0	7.8	11.3
Total energy (%)	100.0	100.0	100.0
Total volume (mboe)	127.4	155.8	219.8
Average annual growth (%)	–	6.9	7.1
Power use (%)	37.5	40.0	41.0
Oil share in power use (%)	48.2	32.1	16.0

Source: Green (1997:87)

and may open in the future (Green, 1997:87). Predicted demand is shown in Table 9.2.

Renewable energy resources

The Philippines has immense potential for renewable resources of different varieties. The estimated potential for geothermal energy is in excess of 4,000 MW, with proven reserves of 1,850 MW. In 1990, 22 per cent of the country's power needs were supplied from geothermal activity, mainly within Luzon and the Visayas. Hydropower is also substantial, with a total potential of some 8,900 MW, of which 2,217 MW had been developed in 1990 (Green, 1997:85).

Table 9.3 indicates the supply of small-scale renewable energy resources in the Philippines in 1996. The majority of supply comes from biomass, particularly fuelwood and agricultural residues such as coconut and bagasse. Table 9.4 indicates the projected future supply of biomass sources in the Philippines.

In 1996, there was very little development of newer renewable energy technologies such as solar, PV and wind. However, the potential

Table 9.3 The Philippines: new and renewable energy contribution by resource, 1996

Resource	Volume (m bbls fuel oil equiv.)		Per cent	
Biomass	72.42		99.89	
Bagasse		9.67		13.35
Coconut residue		11.67		16.11
Ricehull		3.86		5.32
Fuelwood		42.34		58.40
Charcoal		4.81		6.63
Animal Waste		0.01		0.01
Other biomass		0.05		0.07
Others	0.08		0.11	
Industrial waste / Black liquor		0.01		0.01
Solar		0.02		0.02
Wind		0.00		0.00
Microhydro		0.06		0.08
TOTAL	72.50		100.00	

Source: The Philippines Department of Energy (1997:3)

Table 9.4: The Philippines: supply of new and renewable energy sources 1996 – 2025 (in million barrels of fuel oil equivalent)

Biomass	1996	2000	2005	2010	2015	2020	2025
Rice residues	7.26	8.50	10.34	12.58	15.30	18.62	22.66
Coconut residues	18.48	20.01	22.09	24.39	26.93	29.73	32.82
Bagasse	10.99	12.86	15.65	19.04	23.16	28.18	34.28
Fuelwood	80.29	88.15	99.76	113.71	130.50	150.69	175.00
Animal wastes	11.86	12.35	12.97	13.64	14.33	15.06	15.83
Municipal wastes	3.89	4.38	5.03	5.69	6.39	7.11	7.85
Total	132.78	146.24	165.84	189.04	216.61	249.39	288.44

Source: The Philippines Department of Energy (1997:4)

for these is high. The average insolation value in the Philippines is 5 kWh m^{-2} day^{-1}. In Panay, the average wind speed exceeds 15 km hr^{-1} throughout the year, and it exceeds 10 km hr^{-1} in Basco, Batanes, Cuyi, Palawan and Ilio. These are all coastal, island locations. A serious problem with wind energy in the Philippines is he high incidence of tropical cyclones, which may damage wind energy equipment. An average of 19 cyclones may pass through the country each year with wind speeds greater than 200 km hr^{-1} (Green, 1997:90–7). The most far-reaching developments in new and renewable energy technologies may therefore lie in solar/PV and biomass utilization.

Structure and liberalization of the electricity supply industry

Electricity market and institutions

The generation of electricity in the Philippines at present is heavily biased towards coal and oil use. However, this is set to change with government policies to increase the use of indigenous energy resources. Table 9.5 shows the installed capacity for electricity generation by source 1996–2010, and Table 9.6 indicates programmed and planned capacity additions for electricity generation for the same period. The tables indicate that energy demand is expected to triple by 2010, but that the greatest advances in expected generation will come from local natural gas supplies, hydroelectric power and still some imported coal. Imports of oil are expected to stabilize.

Electricity is supplied via three main grid systems in the Philippines. These are the Luzon grid, which comprises 75 per cent of total national generation and 73 per cent of installed capacity (or 6,932.1 MW in 1995). The Mindanao grid is the second largest, representing 14 per cent of

Table 9.5: The Philippines: installed electricity generation capacity 1996–2010 (MW)

Fuel source	1996	2000	2005	2010
Hydro	2,333	2,572	4,071	4,570
Geothermal	1,414	2,014	2,134	2,859
Coal	1,460	3,310	4,210	4,810
Local	610	610	1,210	1,810
Imported	850	2,700	3,000	3,000
Oil	5,349	6,899	5,999	5,999
Gas	0	0	3,000	4,500
Renewable energy	0	0	0	779
Other fuels	0	0	3,850	13,351
Imported coal	0	0	2,300	6,300
Imported oil	0	0	1,550	7,051
TOTAL	10,556	14,795	23,264	36,868

Source: The Philippines Department of Energy (in Lefevre et al, 1997a:79)

Table 9.6: The Philippines: programmed and planned capacity additions 1996–2010 (MW)

Year	Coal	Geothermal	Large hydro	Oil	Natural gas	Other fuels	Total per year
1996	–	220	–	50	–	–	270
1997	–	480	–	100	–	4	584
1998	300	80	–	550	–	10	940
1999	1,550	–	140	–	–	29	1,719
2000	–	40	29	–	900	28	997
2001	300	–	32	–	900	174	1,406
2002	250	–	68	–	600	966	1,884
2003	250	–	–	–	600	471	1,321
2004	100	–	750	–	–	961	1,811
2005	–	120	569	–	–	1,359	2,048
2006	120	80	16	–	1,500	2,297	4,013
2007	120	–	7	–	–	2,297	2,424
2008	120	90	9	–	–	1,797	2,016
2009	120	60	3	–	–	1,797	1,980
2010	120	495	464	–	–	2,092	3,171
TOTAL	3,350	1,665	2,087	700	4,500	11,985	24,487

Source: The Philippines Department of Energy (adapted from Lefevre et al, 1997a:76-7)

national generation and 16 per cent capacity (1,552.2 MW). The Visayas grid is the third smallest, with 10 per cent of generation and 10 per cent of capacity (927.1 MW). The Visayas is the most diverse grid, involving the smaller islands of Bohol, Cebu, Leyte, Negros, Panay and Samar.

In addition to these large grids, there are small island grids on the more remote islands. These contributed just 1.5 per cent of total generation in 1995, or 149.4 MW. Microhydro sources supplied just 1.8 MW to this total, the rest coming from diesel generators. In general, the transmission and distribution of electricity via grids is conducted via transmission lines of between 69 kV and 500 kV. The grid system is being extended (see Table 9.6). Interconnection between islands started in 1983 with the link between Leyte and Samar in the Visayas, followed by connections between the neighbouring islands of Negros and panay in 1990 and then Negros and Cebu in 1993. Connections between the Visayas and Luzon main grids started in 1994, with the long-term aim of unifying all three main grids (Lefevre et al, 1997a:48).

Table 9.7: The Philippines: substation capacity and transmission line length 1990-95

Substation rating (kV)	Substation capacity (MVA)		Transmission line length (ckt-km)	
	1990	1995	1990	1995
500	–	–	490	853
230	7,590	7,640	3,634	4,064
138	3,243	3,619	2,913	3,362
115	2,027	2,147	508	508
69	1,291	1,252	5,760	6,667
Below 69	230	230	755	755
Total	14,381	14,888	14,060	16,208

Source: The Philippines National Power Corporation (in Lefevre et al, 1997a:48)

The main electricity utility is the National Power Corporation (NPC, or 'Napacor'). The NPC was established in 1971 under the Republic Act (RA) 6395. This was then amended under the Presidential Decree (PD) No. 40, which established NPC as the sole generator of electricity in the Philippines. In 1987 this monopoly was broken in principle by Executive Order (EO) 215, which started the basis of privatization (see next section).

The second major institution of electricity supply in the Philippines is Meralco, or the Manila Electricity Supply Company. Meralco gained its name when it was founded in 1903 as the Manila Electric Rail and Light Company, and it is today the largest privately owned distribution utility in the Philippines, with a franchise area of 9,328 km^2 (or 3 per cent of national territory). In 1995, it supplied 14 cities and 97 municipalities within this area, including 2.7 million customers. Its sales in 1995 were 12,279 GWh, or 60 per cent of the total electricity supply nationally (and 78 per cent of Luzon). Meralco is 20 times bigger than the next largest distribution company.

In addition to these large utilities, there were in 1995 some 27 private utilities throughout the Philippines which have come into existence since privatization began in 1987. Furthermore, there were 109 electricity cooperatives (ECs), usually in rural or remote locations, which also undertake local generation or distribution of electricity. These local organizations are discussed in more detail in following sessions.

Electricity supply in the Philippines is regulated by three main organizations. The Department of Energy (DOE) is the central planning agency which oversees development and investment. The Energy Regulatory Board (ERB) is the latest body to undertake the regulation of power tariffs in all utilities and corporations, and until oil deregulation it also had responsibility for controlling domestic petroleum prices. The ERB is an independent agency under the Office of the President rather than within the DOE. Precursors to the ERB included the Oil Industry Commission. This was established in 1971 but replaced in 1977 by the Board of Energy when the Ministry of Energy was created.

Finally, the National Electrification Administration (NEA) was set up by PD 269 in 1973 as an organization seeking to increase electrification. The main achievement of the NEA has been to create ECs and supply these with technical, financial and managerial assistance. The NEA traditionally set the prices for ECs, but in 1992 RA 7638 gave this responsibility to the ERB.

Privatization and liberalization

Electricity supply in the Philippines had always been comparatively expensive because of the need to import fuel. Indeed, retail power tariffs have been among the highest in Asia. However in addition to this, some other problems were experienced as a result of the centralized structure of the electricity industry, and the difficulties this posed for supplying a large and diverse archipelago.

First, electricity supply has always been uncertain and subject to delays. For example, during Summer 1993, Metro Manila experienced power outages of up to 10 to 12 hours a day. Such shortages were partly a result of the deferral of a 625 MW nuclear power station commissioned during the Marcos years without existing alternative power sources. In addition, the system generally lacked technical support to implement maintenance or planned baseload expansions on time (Lefevre et al, 1997a:89).

Secondly, the establishment of three major grids through one major utility (NPC) had led to cross subsidies which made the system economically inefficient. For example, in 1995 the cost of the small island grids was estimated to be 3.5 Pesos (P) kWh^{-1}. The tariff charged was P1.75 kWh^{-1}, hence making the subsidy P1.75 kWh^{-1}. This compared with a cost of P2.8 kWh^{-1} for the Visayas grid (plus tariff and subsidy of P2.0 and P0.8 kWh^{-1} respectively); and P1.5 kWh^{-1} (tariff P1.3, subsidy P0.2 kWh^{-1}) for Mindanao. The source of these subsidies was Luzon, where costs were P1.7 kWh^{-1}, tariffs P1.9 kWh^{-1} and therefore extracted value being P0.2 kWh^{-1} (Lefevre et al, 1997a:121). This cross subsidization also affected different electricity uses: for Meralco in April 1995, residential customers received P0.85 kWh^{-1} and street lighting P2.01 kWh^{-1} while commercial and industrial customers contributed P0.38 and P0.39 kWh^{-1} each respectively.

As a result of long-term problems like these, the Philippine government undertook a programme of privatization and liberalization to improve the efficiency and speed of electricity generation. Two key laws were influential in liberalizing power (see Lefevre et al, 1997a:89, 103).

EO 215 of 1987 provided the legal framework for private investment in power generation. This order amended PD 40, which gave the NPC sole right to generate power in the Philippines. In 1989 the Implementing Rules and Regulations (IRR) of EO 215 were published. The IRR still maintained the position of the NPC as the overriding authority on strategic electricity supply topics and grid advancement. Moreover, the IRR limited IPP activity to cogeneration and electricity plants consistent with NPC ambitions. The IRR also insisted that all IPPs under the so-called NPC power development programme sell only to the NPC after long price negotiations. Fortunately, the so-called DOE Law (RA 7638) in 1992 relaxed the IRR in order to allow IPPs to sell directly to electricity distributors and so accelerate the provision of power. Further revisions to the IRR in 1995 transferred full accreditation powers for new IPPs from the NPC to the DOE, and identified four key areas of new investment including cogeneration; renewable

Off-grid renewable energy under active investment

energy development; indigenous energy resources; and grid-supplied power generation.

The second key legislation was the so-called BOT Law (RA 6957) which was approved in July 1990. This expanded the principles of EO 215 to all infrastructure projects in the Philippines and gave a variety of ways for private investors to be involved in activities previously undertaken solely by state agencies.

In addition to these laws, a variety of less important legislation has increased private access to energy markets in the Philippines. In 1991, the Foreign Investment Act (RA 7042) included a variety of incentives to encourage foreign investment. Incentives included features such as income tax holidays, abolition of duties on the importation of capital equipment and various other tax credits and foreign exchange permits. In 1995, RA 7718 amended RA 6957 to expand the types of contracts and schemes available to investors. In the Philippines Power Summit in October 1995, President Ramos also directed the NPC and the DOE to eliminate cross subsidies. Future subsidies for socially-oriented electricity use such as street lighting would come from the reduced costs of using indigenous rather than imported resources.

The oil industry was also deregulated in the Philippines in two stages. First, in 1995 the DOE Administrative Order (AO) 95-001 relaxed regulations for petroleum products distribution and marketing, and abolished the need to hold public hearings for expansion or changes to refineries or gasoline and liquefied petroleum gas (LPG) distribution. Secondly, in April 1996, the Oil Downstream Industry Deregulation Act (RA 8180) enforced these changes and introduced full market control on the pricing of oil products.

This programme of privatization and liberalization achieved impressive results. The first IPP in the Philippines was commissioned in January 1991, and was a 210 MW gas turbine in Navotas, Metro Manila undertaken by Hopewell on BOT terms. In 1992, a new fast-track power programme was introduced, leading to the commissioning of 11 IPP projects alone in 1993 (or 1,300 MW). Table 9.7 shows the extent of private power capacity in the Philippines in 1995 according the variety of schemes adopted.

Table 9.8: Private power capacity in the Philippines, 1995 (MW)

Type of investment route	MW
Owned and operated by NPC	5,715
BOT (Build-Operate-Transfer)	836
BTO (Build-Transfer-Own)	836
ROM (Rehabilitate, Operate, Maintain)	1,124
ROL (Rehabilitate, Operate, Lease)	175
OL (Operate and Lease)	400
PPA (Power Purchasing Agreement: an agreement in which the IPP sells only part of its output to the NPC)	262
SIG (Small Island Grids, owned and operated by NPC)	149
Meco IPP (an IPP selling wholly or partly to Meralco)	166

S ource: NPC (in Lefevre et al, 1997a:91)

These new additions to generation capacity have been claimed to have reduced the power shortages of the Philippines. This success and speed of privatization has been attributed to the incentives offered. For example, the NPC offered to take on the risk of investment by guaranteeing fuel supplies or short construction periods. Furthermore, at the time of the BOT law, there were few opportunities for BOTs outside the Philippines, and the recession and low interest rates in many alternative countries made the Philippines more attractive as an investment. By September 1997, there were 30 IPP projects with over 5,000 MW finished in the country, plus another six projects (1,370 MW) under implementation (Lefevre et al, 1997a:92). Under the current government Power Development Plan, only 9 per cent of targeted capacity additions between 1996 and 2010 (i.e. 2,424 MW) are to be implemented by NPC. The rest (more than 24,000 MW) is expected to be supplied by IPPs, including Meralco.

The intended privatization programme, however, is still incomplete and during 1998 the government is still waiting to pass the so-called 'Omnibus Electric Power Industry Act'. This legislation, composed of two bills before both the Senate and the House of Representatives, aims to achieve full privatization of NPC's $4-5 billion assets, and achieve full competition and eventual possibilities for flexible transmission access and wheeling supply to consumers. It is proposed that responsibility for transmission would be assumed by an independent national transmission company (NTC), although the different bills propose alternative restrictions on the extent to which the NTC would be independent from electricity generators. A new agency, the Privatization and Restructuring External Office has been created within the NPC to prepare for full privatization, and 12 potential companies have been identified according to geographical location as a future decentralized model for the NPC.

Renewable energy and technology transfer

The Philippines may therefore be defined as a location in which there are several opportunities for using renewable energy technology because of the existence of local resources and a topography of small islands that make the extension of grids from centralized power production costly and difficult. In addition, the government has undertaken a number of initiatives to encourage foreign investment in electricity generation, and has also highlighted the potential role of renewable energy. This section identifies the implications of these changes for investment in renewable energy technologies and the transfer of environmentally friendly technology resulting from investment.

Vertical transfer

Vertical transfer of renewable energy and technology transfer refers to the direct introduction of new technologies into different locations through investment. This is the first stage of developing renewable energy power generation, and may also lead onto longer-term horizontal integration or sharing of expertise with local communities and companies. Vertical integration may be achieved through foreign investment, but it may also take place through investment by national producers in regions where they have previously been inactive.

In the Philippines, development of renewable energy has been encouraged as part of the governmentÕs scheme to maximize use of indigenous resources. In particular, these have focused on large-scale renewable projects such as geothermal and large hydro. The New and

Renewable Energy Program is being implemented by the DOE though its Non-Conventional Energy Division. This has four strategic sub-programs to address technology development, commercialisation, promotion and localized implementation. In addition, the Philippine Council for Industry and Energy Research and Development of Science and Technology has sought to integrate science research on renewables with practical business development.

The NPC has targeted the construction of 2,859 MW geothermal capacity between 1996 and 2010, including a variety of privately-owned and operated projects. These include the Tongonan power plant operated by PNOC-EDC under a BOO scheme, and a two stage BOT project in Mindanao involving a consortium of Marubeni and Oxbow Power Corporation.

Table 9.9 indicates large hydro projects planned between 1996 and 2005, and Table 9.10 shows the plans for small hydro projects. All large hydro projects except for Kalayaan are offered for private-sector development. In general, small hydro schemes are expected to add 420 MW to the system by 2010, and mini hydro (of up to 10 MW) have been targeted by the DOE to supply 150 MW between 1996 and 2005 (Lefevre *et al*, 1997a:99).

Table 9.9: The Philippines: planned hydro projects 1996–2005

Project name	Grid	Capacity (MW)	Commissioning date
Casecnan	Luzon	140	1999
Kalayaan	Luzon	300	2004
Bulanog Batang	Mindanao	150	2004
Pulangi V	Mindanao	300	2004
San Roque	Luzon	345	2005
Agus III	Mindanao	224	2005
Timbaban	Visayas	29	2000
Villasiga	Visayas	32	2001
Tagoloan	Mindanao	68	2002
Mini-hydro projects	Various	150	1996–2005

Source: NPC (in Lefevre et al, 1997a:100)

Smaller renewable energy technologies such as PV, biomass and wind are only expected to be 4 per cent of planned additions to installed

Table 9.10: The Philippines: small hydro projects under the BOT programme

Project	Location	Installed capacity (MW)	Annual energy (GWh)	Project cost (US$m)
LUZON				
Amburayan A	Benguet	133.8	1,124.8	45.65
C		29.6	90.5	39.03
Pasil B	Kalinga-Apayao	20.0	78.9	126.9
C		22.0	82.5	31.40
D		17.0	65.8	28.50
Saltan B	Kalinga-Apayao	24.0	110.6	29.25
Tinglayan B	Kalinga-Apayao	21.0	53.3	25.00
Tanadan D	Kalinga-Apayao	27.0	79.7	46.6
SUBTOTAL		194.4	686.1	272.33
VISAYAS				
Villasiga	Antique	32.3	92.0	39.28
Timbaban	Aklan	29.1	125.0	42.27
Bongabong	Oriental Mindoro	28.0	160.0	42.27
Aglubang	Occidental Mindoro	13.6	39.0	28.50
Sicopong	Negros Occidental	17.8	30.6	19.28
SUBTOTAL		120.8	446.6	171.60
MINDANAO				
Cateel E	Davao Oriental	17.5	78.7	35.78
Lanon	South Cotabato	21.2	56.3	19.40
Tran AB	Maguindanao	30.6	85.0	56.48
Suwawan	Davao City	18.3	46.6	29.58
Tamugan	Davao City	18.9	78.2	31.17
SUBTOTAL		106.5	344.8	172.4
TOTAL		421.7	1,477.5	616.33

Note: 1US$ = 24 Pesos

Source: Coordinating Council of the Philippines Assistance Programme, Philippines Infrastructure Privatization Programme (February 1995) (in Lefevre et al, 1997a:101-1)

capacity in the Philippines 1996 to 2005, but this amounts to 4,000 MW. Under the Renewable Energy and Power Program (starting 1990), an initial fund of P750 million (US$30 million) has been made available to support biomass and mini-hydro projects through low interest loans. The fund is sourced from the Government Service Insurance System and Social Security System, and is administered by the Philippines

National Bank, Development Bank of the Philippines and the Land Bank of the Philippines. The Development Bank of the Philippines has a facility in order to make loans specifically to solar power ventures at the level of villages. This combines with tax incentives included in RA 7369 (section II 2a) which exempts import taxes for power generators' utilities and accessories.

In September 1991, the Mini-hydropower Incentives Act (RA 7156) provided a number of incentives to accelerate investment in minihydro schemes. These incentives included special tax rates referring to duty-free importation of machinery and equipment; crediting for VAT payable on machinery and equipment; and a seven-year income tax holiday. Furthermore, the aforementioned DOE Law (RA 7638) of 1992 also prioritized renewable investment as one of four targets for foreign investment. These original measures were strengthened in 1996 by the Investment Priorities Plan which included more tax incentives and a simplification of customs procedures for investors in a variety of new technologies including renewable energy.

However, the encouragement of investment in renewable energy technologies by the government has not simply been seen as an increase in private ownership. The Small Power Utilities Group has been proposed as one of the 12 private companies to be spun out of a future privatized NPC, and this new company would be largely responsible for small island grids (SIGs). However, rather than being a totally independent private company, it has been proposed to make this group an independent government agency attached to the DOE. The reasoning behind this is to maintain some government stewardship over rural electrification and currently non-commercially viable power uses. The funding for the new agency would come from electricity sales to end users; a customer benefit charge set up by the ERB; direct subsidies from Congress; and proceeds from government shareholdings in privatized electricity generation assets (Lefevre *et al,* 1997a:118). In essence, this implies that the government is still willing to consider cross subsidization in order to maintain SIGs and small-scale renewable energy technology.

In addition to these early measures to build investment in renewable energy, the Philippine government in 1997 announced a new programme of 'Pole-Vaulting' for the energy sector. The aim of this is to accelerate energy development by focusing on new technologies such as ocean, solar and wind energy in the short term (the 'OSW' strategy), rather than see these as accessible only after years of investment in fossil fuel technology. The strategy is to be conducted as part of the Thirty-year National Philippine Energy Plan from 1995 to 2025.

The Pole-Vaulting plan involves aspects of both vertical and horizontal integration, as it includes inviting foreign investors in high technology, as well as government initiatives to build capacity and local expertise. The short-term cost of plans from 1997 to 2001 is estimated at

---

Box 9.1: The Philippines 'Pole-Vaulting' scheme

In 1997 the Philippines government announced the Pole-Vaulting task force as a way to accelerate investment in indigenous renewable energy resources. It is an example of vertical technology transfer because new progress will be sought by external investment and rapid deployment of new resources, although there will also be investment in local dissemination and institutions for monitoring resources.

The focus of the scheme is on ocean, solar and wind technology (OSW), where it is estimated that there are a potential 266 million MW of supply, compared with the Philippines anticipated energy demand in 2,025 of 250,000 MW. The aim of the programme is to reduce the comparative high cost of renewable energy by increasing the scale of production.

Specific plans include the identification of key sites for renewable energy development, and the creation of a Renewable Energy Application and Development (READ) Centre to implement new development of energy, and establish solar manufacturing sites at a variety of locations throughout the country. In addition, the centre will undertake mapping of wind resources in order to identify where wind energy may be successfully harnessed.

Wave energy measurements are planned in the Babuyan islands in Batanes; the north-east section of Cagayan; the Pacific costs of Camarines, Sur, Sorsogon and Catanduances; the eastern coast of Samar, Davao Oriental, Infanta, Bolinao and the northeast of Ilocos Norte. Private investment will be sought through cooperation with Asia Power International and NPC for a 1 MW tidal current plant.

In time, it is hoped that the Pole-Vaulting scheme will result in the construction of grid-supplied renewable energy, plus the supply of local manufacturing factories from this energy source.

Source: DOE Energy Plan and documentation, 1997

US$30 million. Initial plans are to prepare specific sites for development and utilization of OSW, and the establishment of a Renewable Energy Application and Development (READ) Centre. In time, it is hoped the centre will increase the mapping of wind resources, implement the planned power systems for each location and promote the establishment of solar manufacturing plants (see Box 9.1).

In addition to Pole-Vaulting in general, the Philippines is the sight of the world largest PV dissemination project to date. This project, the Municipal Solar Infrastructure Project (MSIP), represents the large-scale vertical transfer of technology by British Petroleum (BP) Australia, but is also supported by the Australian Export Finance and Insurance Corporation and the Australian aid agency AusAID (see Box 9.2). The case study shows the way in which international investment may be integrated with the local development and energy supply strategies of the government.

Horizontal transfer

Under the Mini-hydropower Incentives Act (RA 7156) of 1991, horizontal technology transfer is encouraged by tax credits available to foreign investors when using locally manufactured machinery, equipment

---

Box 9.2: The Philippines Municipal Solar Infrastructure Project

The Municipal Solar Infrastructure Project (MSIP) is to be undertaken by BP Solar (Australia) in conjunction with the Australian Export Finance and Insurance Corporation, and the Australian aid agency AusAID. It represents an example of vertical technology transfer by allowing imported PV technology to be integrated into local electricity supply projects.

The aim is to design, supply and install 1,003 solar powered electricity generators for more than 400 villages in Mindanao and the Visayas in the southern Philippines at a total cost of some US$36 million. The PV systems will supply a variety of applications such as street lighting, water pumps and vaccine refrigerators, to larger requirements including regional hospitals with a 4.5 kWh daily load and both AC and DC outlets. Solar thermal systems will also be used alongside PV.

Source: BP Solar promotional material

and materials relevant to hydro power instead of imported plant. Under this arrangement, a tax credit equal to 100 per cent of the amount of VAT and customs duties which would have been paid on imported goods is made available to companies when they buy locally. This incentive has the effect of encouraging demand for locally produced goods and services relevant to mini-hydro power (Lefevre et al, 1997a:103).

The DOE Law (RA 7638) of 1992 also included instructions under its IRR (section 5i) to boost the power and non-power related benefits to communities hosting electricity projects. The IRR required producers to set aside P0.0025 per kWh of total electricity sales as a fund for electrification of host communities. The IRR also stated that host communities should get priority supply at times of power outages, plus priority in employment opportunities. Further money should also be directed at development and livelihood issues locally.

Direct investment in renewable energy has been integrated with general plans for rural electrification through the activities of local electricity cooperatives. In 1995, total sales from ECs amounted to 4,315 GWh, or 16 per cent of national sales. These focused mainly on barangays, or village-based organizations. The Philippines Energy Plan aimed to electrify all municipalities within EC franchises by 1996, all barangays by 2010 and all potential customers by 2018. In addition, the ECs are expected to rehabilitate 104,472 km of distribution lines and expand the

Table 9.11: The Philippines: performance of electricity cooperatives 1990–95

	1990	1995	% increase or decrease
No. of ECs	119	119	–
Status of energization:			
No. of connections	3.2	3.9	4.04
No. of municipalities	1,301	1,365	0.96
No. of barangays	21,314	23,610	2.07
Electrification level* (%)	54	55	0.37
System losses (%)	21.9	18.2	(3.62)
GWh sales	2,883	4,315	8.34

Note: * number of actual versus potential connections

Source: DOE and NEA (in Lefevre et al, 1997a:65)

existing system by 105,825 km and install a new substation capacity of 3,185 MVA within the course of the plan (Lefevre *et al*, 1997a:85). Table 9.10 indicates the performance of the ECs at the local level in the Philippines as an indication of decentralization of electricity generation and distribution. Renewable energy has been identified as an important part of reaching these targets.

The plans for rural expansion have clearly indicated a need for new technological hardware, but in addition a need for new skills and training. The government has therefore started a Performance Improvement Programme as a way to ensure better all-round achievement of electrification aims. The cost of the NEA electrification programme has been put at P104 billion (US$4.3 billion) for the project until 2025, but this is only 1.35 per cent of all total expenditure on forthcoming electricity generation.

These measures so far have largely been government dominated. In addition, the Philippines has also encouraged the adoption of renewable energy systems through allowing non-governmental organizations. These organizations include the Affiliated Non-Conventional Energy Centres (ANECs) mostly comprised of 20 state universities in different regions, and the University of Philippines Solar Laboratory. The Solar Laboratory has prepared and submitted to the government the necessary standards and test procedures for PV equipment. The Renewable Energy Association of the Philippines has also been established to represent private manufacturers and developers.

Some local institutions have collaborated with international donors in order to develop renewable energy. For example, the Phil–German Photovoltaic Water Pumping Project through the University of San Carlos had installed 14 pumps by September 1997 in remote villages in Cebu, Mindoro Occidental, Cavite and Leyte. Local rural water and sanitation associations were also organized to manage these systems. The project is linked to the Special Energy Programme of the German government, which has set up a development project in the Philippines known as Isang Libong Bahay: Pailaw Mula Sa Araw (or Prosolar for short). The programme aims to provide commercial incentives to private manufacturers and sellers to offset pre-commercialisation expenses. As a result of this, some 25 SHS units were installed in Isla Verda and Tingloy Island in Batangas province (NCED, 1997:6–8).

In addition, international non-governmental organizations are also active. The Renewable Energy Support Office (REPSO) of Winrock International, a US non-profit making development organization has helped facilitate renewable energy development through providing training, expertise and institutional support to investors and local implementers. REPSO is implemented locally through a coordinating organization, Preferred Energy Investments (PEI) in Manila. PEI has already identified, developed and funded a variety of projects including the 7 MW grid-connected Bubunawan mini-hydro in Mukidnon province, Mindanao, and the 98 kW Villa Escudero micro-hydro project at an ecotourism site in southern Luzon. In addition, PEI has collaborated with the Philippines' pioneer PV company, the Solar Electric Company, for disseminating PV systems in six locations across the country.

The activities of PEI have also ensured greater success of international investors and greater dissemination of technology and expertise to local populations. The example in Box 9.3 shows how an international investor in biogas power generation has been able to integrate with local communities and local environmental policy. The problem of urban and agricultural waste in the Philippines has been growing with the size of the cities, and at times this has led to conflict because of alleged water pollution. The creation of new biogas generators that use waste allows international investors to increase their operations in the Philippines as well as partially remove the problem of waste. PEI was instrumental in bringing about this investment by acting as an intermediary between local end users and factories, and the biogas investor.

Summary and conclusions

The Philippines is the most active of the four case studies in the building of renewable energy through international investment. This is the result of an excellent physical potential for renewable energy technologies because of the country's islands, volcanoes and availability

> **Box 9.3 Power from waste in the Philippines**
>
> Foreign investment in urban and agricultural waste management in the Philippines makes a good example of public–private synergy in environmental policy. Silk Roads, Ltd. of San Francisco has invested in biogas anaerobic digestion plants in the locality of Ayala Alabang, 30 km south of Pasig City in greater Manila. The plant uses animal waste from pigs and cows in local farms, and can be adjusted to use human sewage too.
>
> Silk Roads has constructed the biogas digester plant, but works in collaboration with the local authority and a women's NGO who are responsible for waste collection. It is not in Silk Roads' interest to increase costs and expertise to collect waste, as this represents additional investment costs with potential low commercial return. However, it is in the urban authorities' interest to dispose of waste. Increasingly, the growth of new pig farms on the urban fringes is leading to conflicts concerning water pollution. Using this waste for power generation helps remove the waste, and contribute to local, grid-connected power needs.
>
> The key requirements for this project are a PPA in order to guarantee income to Silk Roads, plus cooperative local partners with similar aims. The company had already tried to create a similar project in Thailand, in the far northern province of Chiang Rai, but failed to receive a firm offer of a PPA from the local electricity authority. This problem did not exist in Manila.
>
> Silk Roads has established a new locally registered company, Philippines Biosciences, in which it holds 50 per cent of the shares, as a way to consolidate on its position in the Philippines. A partner in this project is Preferred Energy Investments, the specialist intermediary organization for energy development in Manila. The new company aims to establish 69 new biogas sites in the Philippines by 2000. This complementary relationship between Silk Roads and local authorities and citizens groups is a way of integrating international investment in technology into existing local economic and organizational structures, and therefore extending the adoption of the technology.
>
> *Source: Silk Roads Ltd and PEI, 1997*

of biomass resources, plus a government policy that has deliberately sought to increase international and domestic investment in renewable energy. In addition, the country has long-standing trading and language links with Europe and North America, which aids investment.

There has been much direct, vertical integration of new technology into the Philippines in PV, wind and biomass development. The MSIP is an example of a project where international technology is used in order to supply decentralized electrification. In addition to this, the government's Pole-Vaulting programme is an ambitious plan to develop indigenous new renewable energy technologies. However, horizontal integration has also been necessary both in seeking local partners to

Off-grid renewable energy under active investment 185

ensure the success of biomass investment, as in the case of Silk Roads, or in facilitating the long-term education in technology in rural electrification schemes.

Political will and organization, however, remain important problems. The Omnibus Electricity Bill may radically reform the private-sector involvement in renewable energy and other forms of power generation. But these measures have to be implemented. The government has stated that it is willing to invest large sums in complex technology such as PV and ocean thermal generation, despite the evidence from other parts of Asia that these schemes are costly and require long-term commitment. Many past schemes in the Philippines have suffered from bureaucracy and corruption. In September 1997, the parliamentary energy committee recommended President Ramos to revamp the NEA be revamped following an investigation into alleged graft in implementing its rural electrification program. The committee had alleged that the program had been delayed by favouritism in the allocation of contracts and an inefficient use of funds (*Philippine Energy Digest*, 2:17, September 1997, p1).

Privatization has clearly opened up investment opportunities in renewable energy in the Philippines, and the country remains attractive to foreign investors in comparison with the more rigidly controlled neighbours of Vietnam or Indonesia. However, the role of government is still important for large-scale renewable energy development in the Philippines, even if it is not always efficient. Intermediary organizations such as PEI, and the existence of local NGOs and electric cooperatives in rural zones, present complementary mechanisms to allow investment to go ahead. The highly developed nature of these additional institutions to central government present a positive model for enhancing international investment and technology transfer for renewable energy.

Part III
Conclusions

Chapter 10

Renewable energy investment and technology transfer in South-east Asia

Introduction

Chapters 10 to 12 form the conclusions to the book. Chapter 10 summarizes the case studies in Part II, and draws lessons for building renewable energy investment and technology transfer based upon the experience of South-east Asia. Chapter 11 presents the implications of these lessons for the climate change negotiations, and in particular the debates concerning international investment and technology transfer. In Chapter 12, the book concludes by assessing the use of business investment in global environmental policy in general, with implications for business regulation, privatization, and national technology policies.

This first concluding chapter focuses on the empirical case studies of renewable energy investment and technology transfer in South-east Asia. A main objective of the book has been to assess how market forces and technological development may be combined to extend renewable energy to new locations. This chapter addresses this objective by considering the following questions:

- What market structures in the case studies of Asia have allowed the maximum investment in renewable energy technology?
- What commercial and institutional steps are needed to ensure that new businesses succeed?
- How may technology transfer be interpreted and maximized for renewable energy?
- Which technologies present best options for business growth and local adoption?
- What policy lessons can be extracted for other locations undergoing rapid industrialization and privatization?

The chapter is divided into four main sections. Firstly, the evidence from the South-east Asian case studies is summarized in order to draw conclusions about the impact of different market structures and regulatory regimes upon business and technological development. Secondly, the chapter then looks in more detail at renewable energy business development in terms of technological choice and funding schemes. Next, the chapter focuses on technology transfer and the implications for ensuring the success of both vertical and horizontal integration. Finally, the chapter lists some policy recommendations for maximizing renewable energy investment and technology transfer under conditions of rapid industrialization and privatization.

Investment trends and structures in South-east Asia

The aim of the case studies in Chapters 6 to 9 was to identify the impact of different business and market structures on renewable energy technology investment and transfer. The four structures (in Table 5.1) were identified in order to describe different reliances on grid-supplied electricity, competition from fossil fuels, bureaucratic control of investment and vertical or horizontal integration of investment. The broad findings observed in each category are summarized in Table 10.1.

It is acknowledged that each of the four case studies presented Ð of Thailand, Vietnam, Indonesia and the Philippines Ð may vary from the descriptions of each category at the sub-national scale. Similarly, the privatization of the electricity supply industries in these countries is still ongoing, and relatively poorly developed in comparison with Europe or North America. However, some conclusions may be drawn to illustrate the nature of investment in renewable energy, and what steps are needed to maximize commercially successful technology transfer in the long term at both the national policy level, and at the local project scale.

In Thailand, the dominant presence of grid-supplied electricity has inevitably meant that large-scale rural electrification schemes based on decentralized renewable energy technology are not feasible. However, the most eye-catching policy measure has been the SPP legislation, which has allowed factories involved with agricultural or urban waste to adopt cogeneration technology, and sell the surplus to the national grid. This measure may originally have been part of energy saving, or DSM policies, but they have also had the effect of attracting private investment into renewable energy. Other, generally off-grid, technologies have been encouraged by government agencies such as the Telephone Organization of Thailand, or international agencies like CASE, which have provided pilot projects to demonstrate the feasibility of such technology.

Table 10.1: Adaptation of Table 5.1 to summarize characteristics of investment and technology transfer in the case study categories

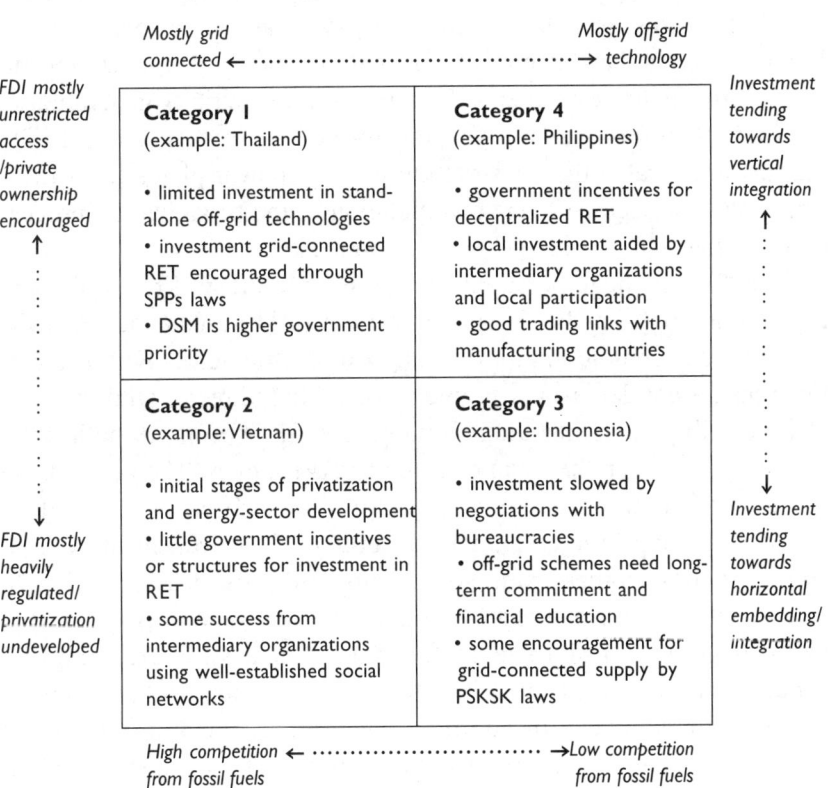

	Mostly grid connected ← · → Mostly off-grid technology		
FDI mostly unrestricted access /private ownership encouraged ↑ ⋮ ⋮ ⋮ ⋮ ⋮ ⋮ ⋮ ⋮ ↓ FDI mostly heavily regulated/ privatization undeveloped	**Category 1** (example: Thailand) • limited investment in stand-alone off-grid technologies • investment grid-connected RET encouraged through SPPs laws • DSM is higher government priority	**Category 4** (example: Philippines) • government incentives for decentralized RET • local investment aided by intermediary organizations and local participation • good trading links with manufacturing countries	Investment tending towards vertical integration ↑ ⋮ ⋮ ⋮ ⋮ ⋮ ⋮ ⋮ ⋮ ↓ Investment tending towards horizontal embedding/ integration
	Category 2 (example: Vietnam) • initial stages of privatization and energy-sector development • little government incentives or structures for investment in RET • some success from intermediary organizations using well-established social networks	**Category 3** (example: Indonesia) • investment slowed by negotiations with bureaucracies • off-grid schemes need long-term commitment and financial education • some encouragement for grid-connected supply by PSKSK laws	
	High competition ← · → Low competition from fossil fuels from fossil fuels		

Source: the author

In Vietnam, there has been little official institutional support for renewable energy at the national level other than large-scale hydro schemes. The key problems have been the general rapidity of change within Vietnam, which has seen investment accelerate faster since the early 1990s than the ability of government to transform in order to manage it. The government is intent on increasing power supply from large hydro or thermal sources. Small-scale projects have been advanced by entrepreneurial agencies such as SELF, who have used existing social networks, such as the Vietnamese WomenÕs Union as a partner. However, investment in small-scale renewable energy projects remains generally small – largely because of the continued bureaucratic problems of gaining access to markets, the poor paying ability of local users in Vietnam and the lack of market-oriented intermediary organizations to help overcome these problems. Privatization if electricity generation in Vietnam, therefore, has not yet transformed the supply of electricity but just accelerated the construction of large power plants from IPPs. There has not been an overt manipulation of the privatization process to accelerate renewable energy development.

In Indonesia, huge potential for investing in off-grid renewable energy technology has been hampered by a rigid bureaucracy and the disparate and remote nature of many rural settlements. Furthermore, the transmission losses and problems of dealing directly with the state electric utility have encouraged many foreign investors to build their own power plants rather than rely on official supplies. The government has adopted grandiose schemes for rural electrification backed by ODA, and using indigenous PV technology. However, most demonstrable success comes from isolated off-grid projects using wind or PV, in which intermediary organization such as Winrock International have established successful management structures and distributed (decentralized) utilities, as well as attending to the technical upkeep of machinery. In addition, a non-governmental network of renewable energy suppliers offers to bypass state bureaucracy. Concerning grid-connected renewable energy, Indonesia has adopted a SPP incentive scheme, the PSKSK, which has encouraged foreign investors to adopt minihydro or biomass-fired plants and sell surplus generation to local grids. However, full enhancement of renewable energy in either rural or industrial locations is still hampered by political uncertainty over long-overdue changes to state bureaucracy, and the physical remoteness in many islands and the imbalance of the population.

In the Philippines, renewable energy investment seems the most advanced because of the physical suitability of many of the islands, and government policies that aim to 'pole-vault' the Philippines' technological capability in renewable energy into the twenty-first century by adopting the latest PV, wind and ocean technology. The practical ability of the government to implement these ambitious plans remains to be seen, but in principle the Philippines offers a variety of lessons for other countries willing to accelerate foreign direct investment into renewable energy. Privatization has allowed a variety of small and renewable-energy oriented investors to expand, although many of these projects are also linked to ODA. The high level of public participation in NGOs and electric cooperatives has also assisted the development of decentralized rural electrification, or the synergistic integration of local waste collection with biomass power plants operated by foreign investors. The creation of the intermediary organization, Preferred Energy Investments in Manila has also allowed foreign investors to overcome bureaucratic controls, and gain closer contact with local markets.

Two important conclusions from this analysis at the national level are:

1. Large grid domination and price competition does not mean that renewable energy technology is uncompetitive. In Thailand, the SPP legislation has shown that factories using biomass may be encouraged to set up their own power plants and sell surplus power to the grid. This success shows that market niches exist for renewable energy development even under apparently difficult market conditions if legislation such as the SPP ruling makes it legal and commercially attractive.

2. Rigid state bureaucracies may be bypassed through the establishment of intermediary organizations willing to represent the interests of foreign investors. Organizations such as Preferred Energy Investments

in the Philippines or the Renewable Energy Network of Indonesia enable investors to identify achievable projects in contact with end users, which can then be monitored over a long time period. These organizations constitute an alternative to standard project financing or BOT structures because they represent a body of expertise in renewable energy finance and maintenance that may be missing in state bureaucracies.

However, there are also two important apparent barriers to renewable energy investment:

1. Renewable energy remains very small in the overall future power supply of South-east Asia. Most energy supply will come from thermal or large hydro schemes, particularly using coal and natural gas. Indonesia is the region's largest exporter of coal and gas, and Vietnam and Thailand also have their reserves. Most governments' priorities are to minimize dependency on imported oil. Energy and environmental policies may therefore be most focused on the replacement of coal with natural gas, or the implementation of DSM measures, into which some renewable energy policies may be integrated (as in the case of the SPPs in Thailand).
2. Political bureaucracies in South-east Asia have still remained resistant to change. In Indonesia and Vietnam state control of investment and energy remains dominant, and in Thailand the state utility, Egat, has resisted attempts to privatize it on the grounds that it is a natural monopoly. Privatization and liberalization does not – of course – imply a total absence or reduction in government regulation. However, some other aspects of government policy, such as subsidizing fossil fuels, or being receptive to decentralized power production inspired by foreign technology and investment, may not be given sufficient emphasis.

These are some key conclusions concerning national trends for renewable energy apparent from the case study chapters. The rest of this chapter now assesses the lessons of the case studies for renewable energy technology and technology transfer in general.

Local renewable energy development

This section of the chapter assesses the factors underlying the development of renewable energy technology in Asia. Two key aspects are identified: technological choice, and investment and funding structures. Aspects of technology transfer are considered in more depth in the final section.

Technological choice

The case studies have indicated that there are a variety of investment opportunities in renewable energy technologies in South-east Asia. The general trend is of a major increase in PV technology, with available imports from UNFCCC Annex I countries such as Japan, the United States, the UK and Germany. NEDO of Japan has stated its support for the 'zero-emission village', or decentralized rural electrification in Asia using PV technology. These futuristic visions have also been supported by the grand schemes within Asia: the 'one million homes rural electrification project' of Indonesia, for example, is largely based on PV technology. The dominant form of PV electrification proposed is stand-alone SHS, or individual PV systems in homes, rather than distributed grid-connected generation (in which remote PV units supply a grid) or centralized grid-connected generation (large banks of PV in one location supplying the grid).

However, there is concern from other technology producers that PV should not be seen as a panacea. Wind technology in particular has the immediate advantage of being linked to water pumping. However, investment in wind has to date been hampered by the absence of reliable wind speed mapping in South-east Asia, and the outdated belief that typhoons may threaten wind turbines (Bergey, pers. comm. October 1997). More research needs to be done into the potential locations of wind turbines, and the various operating and maintenance costs of wind technology.

Furthermore, biomass renewable energy sources are widely available in both agricultural and urban regions, and power generation using biomass has been shown by the case studies to be comparatively easily integrated into local economies (see examples of SPPs in Thailand, and power from waste in the Philippines). There is also potential for developing biomass technology for SouthÐSouth technology transfer, as indicated by the export to Vietnam of Indian biomass generating equipment. Small hydro technology is similar to biomass technology in that it can be integrated into local economies and grid systems by mining or manufacturing companies (see examples of companies in Sumatra, Indonesia). Microhydro systems are also usable in mountain areas for small-scale electricity generation within villages.

Large-scale hydro projects are problematic because they often involve the relocation of villages, and have become a symbol in Thailand and the Philippines of local opposition to the state. Large ocean technology is also problematic because it has yet to be proven cost effective, despite the commitment of the Philippine government to developing it. Geothermal activity, however, does present a way ahead for large-scale grid-connected renewable energy development, if pricing mechanisms can be sought to guarantee investors' risk. In Indonesia, the organization of large hydro and geothermal development by complicated government bodies has slowed foreign investment in these projects because projects are not always evaluated on a market basis, and because ministries have to consider various aspects of development in addition to power generation.

Investment and funding structures

Financial structures for renewable energy investment continue to be complex and evolving (see Kozloff, 1994, 1995a, b, 1998). In the case studies, the main impetus for investment seems to be the combined efforts of entrepreneurial companies (including multinationals) in conjunction with selected use of ODA. One example of this is the BP Solar PV investment in the Philippines.

In addition, much success has been achieved by smaller projects using PV or wind technology administered by specialist intermediary organizations such as Preferred Energy Investments, Winrock International and CASE (see examples of wind in Eastern Indonesia, PV hybrid systems in Thailand and biomass development in the Philippines). These specialist intermediaries have used specialist producers of wind turbines such as Bergey Windpower and World Power Technologies Inc. of the United States, and introduced them to local market conditions.

The traditional model for renewable energy finance is direct project finance, or the identification of funds for specific projects with finite objectives, and where money comes through BOT schemes or bilateral or multilateral investors or donors. However, this had problems for many renewable energy projects because of the complexity of projects undertaken, and increasingly this model is being replaced by more complicated forms of finance. Box 10.1 explains the details of a 'revolving fund', which is one way to integrate local ability to pay with the risk borne by investors in small-scale renewable energy projects.

Box 10.1: New funding options: the revolving fund

A revolving fund is a fund or scheme where the running costs of renewable energy development are met by the continuous return of loan capital with interest. The success or failure of the fund is measured by the extent to which the costs of providing the credit service are recovered.

A project will have two principle sources of funds: initial funding (often from a grant or debt arrangement), and customer payments. These inputs are used to purchase system hardware, cover installation and possibly some of the later operating costs of the project.

The contribution of customer payments is essential to the success of the fund in order to provide inflow of capital and interest to cover credit and operating charges. Operating costs include three distinctive components:
1. The cost of the resources which are leant on to borrowers: these include interest charges and repayment of debt capital adjusted for inflation and currency movements.
2. Programme operating costs such as staff salaries, administration, equipment purchase and depreciation.
3. Lending risk costs associated with bad debts and delayed repayments.

The aim of a revolving fund is to provide a system of finance which allows projects to be funded at market rates, in which technology providers are not left covering unexpected costs. However, it is common for many hidden costs, such as delayed payments or unexpected depreciation of equipment, to be overlooked.

Source: Gregory et al (1997:73-5)

The role of subsidies in renewable energy development is also under review. In general, developing country governments have subsidized indigenous or fossil fuel resources in order to accelerate economic growth. Furthermore, rural electrification has also been subsidized in order to provide electricity quickly to remote areas. However, cross subsidization between urban and rural customers, or between different grids, has been attacked during privatization programmes as evidence of economic inefficiencies (see example of the Philippines). Under privatization, it is not possible to remove all susidy or market distortions, and so the greatest acceleration of renewable energy may come from continued subsidies to renewable energy development but on terms that build investor and consumer confidence in renewable energy. Many past schemes of ODA to renewable energy have undertaken selective or short-term subsidization which has discouraged long-term confidence in the market conditions associated with renewable energy development.

Evidence from the case studies suggests that local financial development is dependent on local support for financing schemes. In Vietnam, the funding agency SELF worked alongside one of the country's oldest community groups, the Vietnam Women's Union, to become more integrated with rural inhabitants. In Indonesia, Winrock International worked with local NGOs and established village-based utilities to distribute electricity from wind turbines. Some general rules for sustainable financing seem to be: gain local support for schemes; identify clear lines of responsibility; use trained staff in administration; identify clear business objectives and risks with financial reporting; integration with local spending ability such as income from crop sales; a variety of borrowers (so natural disasters like crop failures do not affect all sources of income); inclusion of all inflation and interest costs; and an inclusion of a premium to cover all potential technical maintenance costs (Gregory et al, 1997:95–6).

Internationally, funding may also be developing towards communal pools of funds to be used for selective project support, or investment in local stock exchanges. The Clean Development Mechanism might administer a large body of funding from which its Executive Body may seek to support selected projects. For this to be utilized to the maximum, state electricity bodies may establish privatized subsidiaries which are then floated on stock exchanges. Similarly, mutual funds in developed countries specializing in sustainable development projects may also seek ways of supporting renewable energy projects in developing countries. Enhancing the economic structure of government renewable energy projects for easy investment in stock exchanges may therefore increase the flow of funds. An alternative is to direct finance to those intermediary organizations such as Preferred Energy Investments that reduce the non-commercial risks of investors, and which have been shown to be effective in increasing international investment in renewable energy, and in making this commercially and socially sustainable.

Enhancing technology transfer

This section assesses lessons from the case studies for enhancing renewable energy technology transfer. As concluded in Chapters 1 and 4, successful technology transfer is not simply limited to ensuring the technical success of new equipment, but transferring the business and entrepreneurial skills associated with it. This section discusses how success may be achieved for both vertical and horizontal forms of technology transfer.

Vertical transfer

Vertical transfer of technology refers to the successful relocation of technological businesses without necessarily integrating these into local manufacturing. However, evidence from the case studies has indicated that pure point-to-point diffusion may fail without some horizontal integration by training or building local financial capacity. Most advanced technology, such as the PV and wind turbines advanced by CASE in Thailand, Winrock in Indonesia, and SELF in Vietnam, have had problems in either achieving an adequate level of technical maintenance or in attaining commercial and managerial support within new locations. However, careful action by these organizations – often in collaboration with local representatives such as the Vietnam Women's Union – have created support networks that provide these needs.

Evidence for emerging manufacturing joint ventures in the most advanced forms of renewable energy technology is still scant. Some PV joint ventures have been launched in Thailand, plus Indonesia is constructing its own SHS in Java for export to its outer islands. However, at the time of writing, the vertical transfer of technology such as wind turbines and PV is still in general controlled by multinational investors such as BP Solar, who have established subsidiaries throughout the region, rather than seeking to construct this technology jointly with local manufacturers.

Furthermore, the evolution of joint ventures in general has been problematic because of the implications for added cost. The example of the consumer goods giant, Proctor and Gamble, in Vietnam has indicated that enforced joint ventures may discourage foreign investment rather than simply harness it for technology transfer.

Local partnerships have therefore been crucial to the introduction of advanced renewable energy technology in South-east Asia. However, to date these have been characterized by the partnership between mutually complementary organizations such as Bergey Windpower and Winrock, rather than the transfer of knowledge and manufacturing skills from one manufacturer to another. Government agencies have also formed part of these partnerships: in Thailand and Vietnam, for example, departments such as the Telephone Organization of Thailand have performed important roles by adopting PV technology for remote transmission stations as a way of demonstrating and testing the use of equipment. In time, this kind of partnership may lead to an increased market at both government and retail level, for individual usage in villages, factories or new industries such as tourism.

In general, the trend towards the local adoption of international technology supports comments made in Chapter 2 about regional specialization and technology development. In Chapter 2 it was argued that traditional approaches to technology development have argued that competitive advantage comes from indigenous technological development. However, newer approaches (notably from Reich, 1991) have stressed that technology can be developed through international investment and ownership of manufacturing processes, and that this can support alternative local development in a synergistic way. Consequently, the successful vertical transfer of new renewable energy technology such as PV to South-east Asia need not be seen as a failure to gain technology transfer in PV manufacturing. Instead, it should be seen as an opportunity to harness the employment and income (and power supply) generated by the PV manufacturing as a boost to genuinely indigenous manufacturing. Indeed, the case study of PV manufacturing in Brazil (Chapter 3) indicated the pitfalls of attempting to build indigenous PV units without access to international standards and technology.

A key aspect of vertical renewable energy investment revealed in the case studies is that new renewable energy investment has to be seen to be economically justified: either in collecting payments from end users, or in allowing end users to generate income themselves. In this sense, successful long-term renewable energy technology transfer is not achieved through hand-outs, or short-term subsidies as has been the case with some historic ODA packages. However, some funding for public goods such as street lighting and vaccine refrigeration may be provided for by charging a small premium in charges for other uses (as illustrated in the community charge for rural electrification in the Philippines). Grandiose schemes to increase PV usage without also integrating it with local income generation (such as possibly the Indonesian 'one million homes PV rural electrification scheme') may be optimistic and too dependent on aid money. The Asian financial crisis may, however, decrease funds available for importing foreign technologies and hence lead to a renewed dependency on indigenous technology development.

Horizontal transfer

Horizontal integration of technology refers to the education, adoption and eventual manufacturing of new technology by local users. Horizontal integration often follows the successful vertical transfer of technology, and may involve the building of institutional capacity to encourage education and local adoption of new technologies.

One key problem for horizontal integration revealed by the case studies is the fact that most new renewable energy technology such as PV is from the developed world, and has to be integrated into local technical and socio-economic systems. For example, new solar lanterns in rural Vietnam may be damaged because of local fluctuations in power supply. In Indonesia Winrock International had to introduce concepts that water and electricity usage had to be paid for on a metered basis. As noted earlier, the input of PV technology in rural electrification in order to create a 'zero-emission village' is relevant to international concerns about climate change, but not immediately applicable to basic-needs economics of many villages in developing countries.

The tripartite model of local electricity governance advanced by Winrock in Indonesia (see Chapter 8) presents one apparently successful model of vertical and horizontal integration of renewable energy technology. The project involves the establishment of a local electricity utility within villages to administer the distribution of power, but also shares tasks with a local NGO and Winrock itself in order to ensure that full project finance and technical maintenance are observed. Winrock also acts as the liaison with the US-based manufacturer of the wind turbine. The attractive element of this model is that allows the evolution of local financial management at the same time as technical maintenance, without directly going through state bureaucracy. The generation of income by the electricity, and employment of local rather than expatriate staff also imply that the costs of implementing this model are relatively low.

Other forms of horizontal integration have also been achieved in using less advanced technologies, or technologies developed locally. Biomass power development is one key growth area in Asia, and is one where some Asian manufacturers are at competitive advantage over companies from Japan, the United States or Europe because they can be easily integrated into local land use and economic activities. The construction of an Indian biomass power generator BOT project in Vietnam shows that foreign investment may take place from South to South. However, the example of power from waste in the Philippines indicates how it is possible for a foreign investor, with advanced biogas generators, to work well alongside local authorities and pig farmers to achieve synergy of waste management as well as power generation.

A key advantage of using local biomass fuels, and implementing these within locally operate factories, is that they can be encouraged during the process of privatization and liberalization of electricity supply by legislation such as the SPP scheme in Thailand and the PSKSK in Indonesia. These examples show how it is possible to address global concerns of reducing greenhouse gas emissions by appealing to the more immediate local concerns of reducing power bills and letting private investors contribute to grid supply.

Acknowledging local needs, and then integrating these into technology transfer plans, is the best way to ensure that new technologies are adopted into new locations. The motivation for extending renewable energy technology to developing regions may be for many the search for new investment opportunities, and global environmental policy. But the transfer of this technology is unlikely to succeed unless it is acknowledged that many local users will not shares these aims, and that horizontal technological transfer has to be linked to local needs and ability to pay, and international competitive standing. Box 10.2 lists some so-called 'universal critical success factors' for renewable energy investment, identified by one group of researchers (but also see Hurst and Barnett, 1990; Gregory et al, 1997). These indicate that successful technology transfer depends as much on successful vertical integration within market structures, as well as local relevance and persistence.

Conclusion and policy recommendations

This final section draws lessons from the case studies for building renewable energy and technology transfer under conditions of rapid industrialization and privatization. This is done on the basis that privatization and liberalization rarely lead to the total abolition of state interference in energy development, but in the harnessing of private-sector investment for public-sector goals.

> **Box 10.2: Universal critical success factors for renewable energy development**
>
> 1. Investment must fit the medium-term strategy of energy development
> 2. Investment must use proven or reliable designs
> 3. Projects must be based on least-cost approaches
> 4. Appropriate finance must be arranged to cover risks
> 5. There must be adequate marketing and technical staff
> 6. There must be a proven market for the technology
> 7. Do not give free gifts or overt subsidies (such as short-term grants)
> 8. Ensure that a market chain exists between suppliers/consumers
> 9. Consider site specific factors in each location
> 10. Operate in locations where regulations and laws are favourable
> 11. Create an acceptable tariff structure to cover costs
> 12. Disseminate programme results to create market demand
> 13. Conduct adequate project reviews to identify weak points
> 14. Expect demand for products to grow once established
>
> *Source: Stainforth and Staunton (1996), but also see Gregory et al (1997)*

- It is clear from the case studies that foreign investment alone is insufficient to achieve successful renewable energy technology transfer. Renewable energy technology development is not simply a market penetration exercise, or vertical integration in its most direct form. The case studies have indicated that there is also a need for some additional institutional structures in the form of specialist renewable energy intermediary organizations such as Preferred Energy Investments in the Philippines, or the SPP legislation in Thailand, to increase foreign participation in renewable energy development, and lower the costs associated with extending into these new markets. Similarly, renewable energy is clearly not a straightforward commercial choice at present, because it is often justified on the macroeconomic benefits of avoiding ruralÐurban migration and balanced development, plus requires long lead periods with multi-faceted management and technical skills, and also often sells to the poorest consumers in society.
- Liberalization and privatization, therefore, are complementary to renewable energy development in Asia, because they allow the inflow of new technologies (from both North and South), and encourage

decentralized power production with distributed utilities. Yet the acts of privatization and liberalization also have to accompanied by additional policy measures to increase foreign and domestic private involvement in renewable energy development. As Marquand *et al* (1998:16) wrote in relation to renewable energy development: 'regulated privatization may be the only way for countries to achieve short-term and long-term power sector goals, by increasing the efficiency of the existing system and providing financial incentives for the development of the renewables'. In the UK, the NFFO has been universally claimed to have achieved this purpose for renewable energy development (see Grubb *et al*, 1997). But in South-east Asia, similar legislation seems to exist in the SPP scheme of Thailand and the PSKSK of Indonesia. Similar measures exist also in the Philippines in the incorporation of local electric utilities into power development, although this is less specifically for renewable energy development. The SPP and PSKSK legislation may therefore be strengthened or used as a model for other developing countries. The combination of these kinds of laws, and specialist intermediary organizations may also help bypass bureaucratic state structures.

- Renewable energy development depends on this kind of coordinated public-sector 'push' but must also depend on local market 'pull'. Evidence suggests that many renewable energy projects in the past have been built on the principle that there are large proportions of rural populations without electricity. It is important to appreciate that these do not by themselves represent a market, and that market demand represents the identification of particular technologies based on available renewable energy resources and the relative costs of grid extension. As shown in Thailand, the existence of a far-reaching grid does not necessarily imply that there is no opportunity for renewable energy development because the SPP legislation has encouraged grid-connected factories to install their own biomass plants and sell surplus power to the grid. In addition state-operated enterprises such as the Thelephone Organization of Thailand can be used for extending and disseminating PV technology, although this may be an unfair burden to place on state agencies if it

increases costs at the same time as they are undergoing a process of corporatization.
- Privatization of electricity supply industries in Asia, therefore, present a variety of problems of governance as well as economic performance. State operated enterprises are being asked to perform better economically, but at the same time are also being asked to take on public roles. Furthermore, some state departments such as Egat in Thailand are politically conservative in nature, yet powerful enough within the bureaucracy to resist attempts to decentralize their control. It may therefore be better to integrate changes in government bureaucracy with the emergence of alternative agencies for governing the development of environmental or social development. These organizations might include international NGOs or specialist energy agencies. Such transition of social or environmental policies have already occurred in many areas of government policy such as healthcare or biodiversity planning, in a variety of poor, developing countries such as Peru or Vietnam. The implications of internationalization are discussed further in Chapters 11 and 12.
- Renewable energy development, therefore, under conditions of rapid industrialization and privatization may imply an increasing internationalization of both technological supply and governance structures. On one hand, this may boost renewable energy development in countries short of power, and needing rural development. On the other hand, it may present a threat to indigenous technological industries. The case studies have suggested that new renewable energy technologies such as wind turbines and PV are probably best disseminated via a process of vertical integration coming from global suppliers. However, other technologies such as some forms of minihydro and biomass technology can be developed indigenously at better cost and relevance to local needs in developing countries. The increasing internationalization of renewable energy development and governance may then have to be matched by the formulation of national trade and infant industry policies to ensure that local industries are not threatened unduly by this transition. Such national energy technology policies may also include reducing subsidies to fossil fuels –

another long-term requirement for facilitating renewable energy development.
- There is a tendency to present renewable energy exports to developing countries in terms of high technology in remote and poorly developed places. Evidence suggests that stand-alone PV or wind technology can be made to succeed from foreign investment, if this is done in conjunction with cost recovery from end users and integration into local markets. But policymakers must not overlook the ability to use grid-connected biomass, small hydro or geothermal sources that may be utilized using local factories and producers. Such technologies are less visible but may have greater immediate impact on mainstream electricity supply and business practice. The key route to success lies in incorporating part of each stakeholders' aims into each development project: the recovery of costs for investors; the addressing of local villagers' basic needs and desire to generate income; and the requirement of governments to achieve low-cost power supply.

Chapter 11

Redefining international investment and technology transfer for climate change mitigation

Introduction

This chapter presents conclusions concerning the role of foreign investment in technology transfer for climate change mitigation. Two main objectives of the book have been to identify ways to increase the use of foreign investment in climate change mitigation, and to consider the implications of using investment. This chapter addresses these objectives by presenting answers to the following questions:

- How may foreign investment be used to transfer environmentally sound technology for climate change mitigation?
- How may the impasse between North and South in the climate change negotiations concerning the nature and urgency of technology transfer be overcome?
- What are the implications for trade and ownership of technology of increasing the role of foreign investment?
- How may institutional finance organizations act to support greater private investment in climate technology transfer?
- What are the implications for current debates about the CDM?

This chapter is divided into four main sections. The first section assesses implications for technology transfer by reconsidering the meaning of this term under international investment, and how this compares with the current approaches to technology transfer under the UNFCCC. The second section focuses on the implications for foreign investment. In particular, the potential role of technology-related projects in the CDM is discussed, and how this might impact on national or regional development policies. Thirdly, the potential roles for institutional finance are considered. Finally, there is a summary of the chapter's

Redefining international investment and technology transfer 209

main arguments and a list of policy recommendations for governments, investors, and inter- and non-governmental organizations.

It is stressed that the discussion presented here is based on the book's analysis of international investment and renewable energy development. Other climate change mitigating investments such as in natural gas infrastructure, clean-coal technology or energy efficiency (DSM) measures are mentioned, but are not discussed in depth.

Redefining technology transfer for climate change mitigation

The aim of this section is to discuss the approach taken to technology transfer in the UNFCCC and Kyoto Protocol, with the purpose of identifying how this may be improved in order to enhance international transfer of environmentally sound technology. In particular, the section focuses on the differential public–private role for technology transfer.

Vertical or horizontal technology transfer?

Technology transfer is a complex and easily misunderstood term. A common explanation of technology transfer is to see it as a direct sharing of new technology between different parties or countries. In fact, technology is most commonly owned by private-sector investors who are unwilling to share it with potential competitors, and point out that sharing newly-developed technology will eventually discourage any further technology development. As a result, there is much discussion of the importance of technology transfer. But in reality, very little is achieved because of this difference in public wishes and private aims (see Chapters 1 and 3).

An impasse like this has been very apparent in the climate change negotiations. In the UNFCCC, Agenda 21 and Kyoto Protocol, the emphasis on technology transfer has been that it is important for climate change mitigation, but there has been little suggestion of how to achieve it (see Chapter 3 for the wordings of these sections). As discussed in Chapter 3, the reason for this imbalance is that technology transfer has often been used as a bargaining tool, between Northern and Southern countries in the climate change negotiations. In particular, the transfer of environmentally sound technology to countries undergoing rapid industrialization is seen to be one way to ensure that these countries do not significantly increase greenhouse gas emissions in the future. Yet in addition, many developing countries such as India and China have made technology transfer a condition of signing any agreement, and as a counterbalance against unpopular policy suggestions such as emissions trading and JI. It is not surprising that technology transfer is considered to be so controversial and urgent. Yet it is also not surprising that very little progress has been made in technology transfer when it has been dealt with in this emotive way without seeking ways to overcome the public–private impasse.

This book has looked closely at technology transfer in terms of two different models: vertical transfer, or the point-to-point relocation of new technology via investment; and horizontal transfer, the embedding of new technologies via joint ventures, education and management to increase adoption and manufacturing in new locations (see Williams and Gibson, 1990). These terms are also related to the parallel debate in business strategy, in which vertical integration implies the merger of two firms in different stages of production of technology; and horizontal integration, where a company may merge with another in order to increase market share, or form joint ventures and other networks in order to increase access at that stage of production (see Williamson, 1985). It has been argued in this book that the climate change negotiations have focused too closely on horizontal aspects of technology transfer. Horizontal technology transfer is generally seen to be costly and risky to intellectual property by private-sector investors, despite the assurances of the UNFCCC and Kyoto Protocol that these concerns will be taken into account. Increasing technology transfer for climate change mitigation may therefore depend on allowing more opportunities for vertical forms of transfer, or reducing the costs associated with horizontal transfer.

The case studies in Chapters 6 to 9 have provided a variety of examples of vertical technology transfer. PV and wind turbine technologies have been most closely associated with vertical transfer.

Multinational companies such as BP Solar have expanded the supply of PV by opening subsidiaries that have distributed imported technology. However, the most overt examples of vertical integration, such as the BP Solar PV–AusAid project in the Philippines, have still required aspects of horizontal embedding and technology transfer in order to make the new technology technically successful and commercially sustainable. In particular, the activities of supporting organizations such as Preferred Energy Investments in the Philippines, or SELF in Vietnam, have created the commercial and technically-aware circumstances into which much of the imported renewable energy technology has been embedded.

There have also been examples of horizontal integration. The increased supply of power generated from biomass fuel in Thailand, for example, has been encouraged by legislation that makes it commercially attractive for companies to adopt indigenous generation. Furthermore, the extensive efforts of government agencies such as the BPPT, or aid-related organizations such as Winrock International in Indonesia have provided technical support and local political and economic infrastructures for adopting renewable energy technology.

Evidence from the case studies suggests that technology transfer is insufficient when conducted via vertical integration or direct foreign investment alone. However, it is also clear that new, imported technology such as PV and wind turbines, can contribute greatly to local energy development when there are the organizations available to help it become embedded into local technical and economic structures. This suggests that it may be unproductive to identify whether vertical *or* horizontal forms of technology transfer are the most successful, but instead to ask what combination of vertical and horizontal works best.

A successful combination of vertical and horizontal forms of technology transfer, therefore, may be a structure that allows the input of new high technology from multinational investors, but does not threaten the intellectual property rights (IPR) of this technology. In other words, the combination of vertical investment in order to introduce new technologies, and horizontal integration in order to increase technical and financial management. It also implies that technology transfer that aims to increase the number of local companies manufacturing the technology is commercially unattractive to investors and is likely to fail.

Technology transfer for climate change mitigation – in the field of renewable energy development – may therefore be put simply as an increase in the number of people using new technologies, rather than an increase in the number of local companies manufacturing it. This therefore implies that foreign investment must be used to introduce the technology, and that other –possibly public – bodies may undertake the integration of the technology in conjunction with local users and foreign investors. The implications of this for ownership of technology, and local industrial competitive policies are discussed later in this chapter.

Northern or Southern technology?

Another common assumption about technology transfer is that it must always imply transfers of technology from developed, or Northern countries to developing, or Southern countries. As noted again in Chapter 3, this observation overlooks the very great achievements of many developing countries in developing new forms of renewable energy supplies (see Reid and Goldemberg, 1997; TERI, 1997). The case studies in Chapters 6 to 9 also revealed incidences of this: for example, the export of biomass generators from India to Vietnam, or the development of indigenous Indonesian PV technology for the 'one million homes PV rural electrification project'.

It is therefore highly simplistic to make simple dichotomies between the technology from developed countries as complicated and expensive ('circuits and chips'), and technology from the South as cheap and unsophisticated ('sun and dung'). Attempts to increase technology transfer, or to direct foreign investment into technological projects, need to consider that investment for renewable energy development may therefore come from both developed and developing country participants. Furthermore, policymakers need to acknowledge that offering incentives only to investors from those countries which have specific targets for greenhouse gas emissions (Annex I) may have unforeseen impacts on the competitive standing of other renewable energy companies in developing countries.

The impact of international investment upon local industrial competitive standing is a topic of intense uncertainty and debate. Evidence in renewable energy has suggested that single producers of high technology such as PV cannot survive economically if they are not integrated into global networks of research and development, environmental and industrial standards and the upgrading of equipment (see example of the Brazilian PV manufacturer, Chapter 3). This experience does not bode well for the future competitive standing of some high-technology producers in South-east Asia, such as the indigenous PV manufacturing in Indonesia. Yet it does not imply that climate change negotiators or policymakers should assume that these industries should fail, or that they should not be subject to similar incentives of crediting against national greenhouse gas reduction targets as with other companies.

Similarly, the climate change negotiations must not overlook the potential for South–South technology exports. Many have argued that Southern manufacturers are more suited to local technical and financial demands in the South than imports from developing countries (TERI, 1997). Furthermore, the development of rural energy supplies including renewable energy sources is primarily to address basic needs and local income generation, rather than international environmental policy. Some approaches to rural electrification have sought to integrate the local provision of power with adherence to international environmental policy (such as the PV-powered 'zero emission village' proposed by the New Energy and Industrial Technology Development Organization of Japan). Some renewable energy technologies such as PV or wind are indeed more suitable for climate change mitigation than traditional biomass sources because the latter necessarily involve some removal of vegetation and emission of greenhouse gases during generation. But the newer technology may be more difficult to assimilate in rural economies and behavioural patterns than some forms of biomass, and also present a commercial threat to indigenous renewable energy industries.

The use of international climate change policy to justify rural electrification is also politically controversial. Climate change negotiators from developing countries are unlikely to support international efforts to mitigate climate change among the poor rural areas of the developing world without similar actions for other technologies in the developed cities of the North. Furthermore, the inclusion of investment in 'zero emission villages' when existing biomass systems could have achieved the same effect for overall greenhouse gas emissions could create a new form of the 'hot air problem'. This phrase is commonly used in the context of emissions trading to describe the problem when the cumulative reduction in greenhouse gas emissions achieved through trading is equivalent to what would have happened anyway. For international investment, the development of 'zero emission villages' through extension of PV and other sophisticated equipment may be credited through the CDM to the investing country's required reduction in greenhouse gas emissions. However, the net result may be simply to replicate what would have occurred for greenhouse gas emissions with existing biomass generators, plus a loss of competitive advantage for local companies making biomass generators.

Consequently, technology transfer for climate change mitigation needs to be considered in terms of whether this implies a general increase in availability of climate friendly technology (including North to South plus South to South transfers), or simply the increase in commercial opportunities for Northern investors in Southern countries. Increasing North–South technology transfer itself may have unforeseen negative impacts on Southern industrial concerns that are valuable to both local development and climate change mitigation. For some high technology forms of renewable energy, such as PV, evidence suggests that it is impossible for localized producers to maintain competitive advantage against global market forces. However, other forms of technology such as biogas generators may be cheaper and more suited to local conditions than expensive imports, and therefore may allow both the fast implementation of climate change mitigating technology as well as providing business incentives for the South. Integrating technology transfer with international investment must not imply overlooking

Redefining international investment and technology transfer 215

existing successful manufacturers of technology, or assuming that high-technology, environmentally-driven options are most suitable when consumers are driven by basic needs.

Technology or financial management?

Perhaps the most obvious misunderstanding about technology transfer is that it simply refers to the successful integration of equipment into new settings, rather than the overall transfer of management techniques, commercialization and marketing of the equipment. The case studies provided ample evidence that successful technology transfer is not simply 'technical' but also involves the transfer of successful businesses connected to the technology.

This book has made the distinction was made between vertical and horizontal forms of technology transfer. In essence, creating effective management of technology is a form of horizontal integration because it involves building long-term support in the locality for new technology. However, both the technology and the management can be imported directly into the locality, and so effective management may contribute to a shared horizontal and vertical form of technology transfer.

The case studies have suggested that managing technology transfer can be achieved best by combining public- and private-sector actions. In Indonesia, Winrock International successfully implemented wind turbines by building a local management structure. The local management worked well because it was managed by the village utility in conjunction with a regional or national NGO. The facilitating aid organization, Winrock, was available to ensure success but had the ultimate aim of building long-term local financial expertise. Similarly in the Philippines, investment in biogas generation had been accelerated by the careful division of duties between local actors and international investors in terms of commercial advantage of collecting waste and generating power (see the example of Silk Roads). In the rural Philippines, the direct investment by BP Solar in SHSs has been facilitated by the advanced nature of electric cooperatives in villages which have already formed the basis for financial management. In Indonesia, however, the difficulty of achieving a PPA from the government has slowed investment in biomass generation by Bronzeoak.

The lesson of these case studies is that international negotiations over technology transfer need to be aware of the financial implications of successful technology investment. As discussed in Chapter 10, the most viable forms of technology transfer have been those that have incorporated cost recovery as well as aspects of technical management. Enhancing technology transfer for climate change mitigation may therefore imply downgrading the attention given to particular technologies, and instead giving more attention to the management and financial monitoring structures surrounding technology. Increasingly this means allowing full cost recovery locally, but addressing local needs and concerns as a way to ensure this.

Harnessing international investment for climate technology transfer

This section focuses on the practical implications of integrating foreign investment into climate change policy. As noted throughout, foreign investment for climate change mitigation may take various forms, including forestry and land-use projects, as well as energy-related investment such as natural gas, clean coal technology and renewable energy. This section focuses in particular on the ways in which foreign investment may be harnessed for technology transfer.

Opportunities for investment

Negotiations within the UNFCCC on the use of foreign investment for climate change mitigation have faltered because of perceived problems of both developed and developing countries (see Chapters 1 and 2). Classically, developed countries have seen technology transfer as the concern of the private sector and as a costly process that threatens IPR. From the developing country perspective, countries have argued that Northern countries have not honoured their commitments under the UNFCCC and

Redefining international investment and technology transfer 217

Agenda 21, and instead have used foreign investment under JI/AIJ to focus on forestry-related projects that do not aid industrialization.

This book has sought to overcome this impasse by seeking ways to harness foreign investment for technology transfer that do not threaten to raise costs of investment, and which also offer local development for developing countries. It has been argued that this requires a closer look at different forms of both investment and technology transfer, and a greater appreciation that transferring technology is in effect to transfer businesses, with associated commercial standing and financial management. Consequently, successful technology transfer should not just ask 'which technology?', but also 'who should take responsibility for ensuring that local commercial management and embedding takes place?'

The opportunities for harnessing foreign investment for technology transfer therefore depend in part on the existing institutions in different countries for facilitating international investment, and on the regulatory structures that place emphasis on public or private responsibility for technology transfer. Companies with valuable technology are unlikely to invest in countries where transaction costs of market entry and establishing businesses are high. However, there are also differences in the applicability of certain types of technology for vertical technology transfer (such as PV and wind) or horizontal transfer (such as many biomass applications) that have impacts on which forms of foreign investment may be most successful in local economies (see Chapter 10). As a result, harnessing foreign investment for climate change mitigation depends on identifying those sectors of investment that are most likely to succeed in transferring environmentally sound technology to developing countries.

Table 11.1 shows a preliminary classification of different industries with varying potential for technology transfer. In this diagram, categories 1 and 3 refer to those industries where host – or Southern investors – already have a competitive basis, and may provide the basis for future South–South technology transfer. Category 2 is the niche most likely to attract rapid foreign investment via JI or CDM crediting as it refers to technology not currently produced in host countries, but where there is little competitive risk from sharing technology. Category 4 broadly represents the type of technology transfer currently discussed in the climate change negotiations, but is unikely to attract as much investment as category 3 because of the extra costs required in sharing technology.

Table 11.1: Different investment niches for technology transfer

	Expertise and economic base in technology exists locally	Expertise and economic base in technology DOES NOT exist locally
Vertical technology transfer (ownership remains with investor)	1 (associated with high competition and low profit margins)	2 (most attractive to new foreign investors)
Horizontal technology transfer (ownership is shared with local producers)	3 (least attractive to new foreign investors)	4 (associated with high transaction costs and potential loss of competitiveness)

Source: the author

The purpose of this table is to indicate that international technology transfer may be accelerated if it focuses on category 2 where investors may supply new EST to host countries without needing to share it with local producers. In addition, technology transfer may also be achieved via category 4, as currently demanded by climate change negotiators, but only if costs are covered by international organizations such as the GEF; local organizations such as Preferred Energy Investments; or the companies themselves. The division of new investment into these two categories may therefore also identify different responsibilities for accelerating technology transfer. Category 2 indicates those technologies, such as PV, which may be successfully introduced by foreign investors without needing to share with local producers. Category 4, however, indicates technologies such as machinery adhering to ISO 14000 standards is necessary for climate change policy, but for which there are few immediate incentives for either local producers or foreign investors to undertake. In these circumstances, the intervention of international organizations or official development assistance may be beneficial.

However, it is still unclear how new incentives for investment under JI and the CDM may impact on local economic production and competitiveness in host countries. As currently proposed, the CDM may only credit those countries in Annex I for new investments in climate change mitigation. In effect, the CDM will therefore subsidize new imports from Annex I to non-Annex I countries, and therefore reduce the ability of non-Annex I countries to develop their own competitiveness. This problem is now discussed in more detail.

Implications for trade and technology ownership

Table 11.1 suggests that the most effective way to harness international investment for climate change mitigation is in category 2 (vertical transfer) or category 4 (horizontal transfer) when the costs of transfer are borne by other development or aid-related organizations. The case studies in Chapters 6 to 9 support this suggestion with examples of wind and PV investment by multinational companies in the Philippines and Indonesia, with local support and capacity building by specialist development agencies. The evidence of Vietnam suggests also that simple interaction of foreign investors and state organizations has not resulted in progress in either investment or technology transfer, and political struggles emerge to try and maintain businesses on a 100 per cent foreign owned basis.

The implications of these examples are that international investment for climate change mitigation would be accelerated if foreign-owned technologies such as PV could be introduced by investors and then integrated locally with the assistance of specialist agencies. However, as noted in Chapter 3, this format does not necessarily conform with the traditional approach to technology transfer in which manufacturing technology is communicated to local companies. Indeed, the encouragement of investment in new forms of renewable energy technology may threaten existing indigenous industries.

Under newer approaches to technology transfer, however, this format need not be so problematic. As discussed in Chapter 2, newer approaches to technology (notably Dunning, 1993, 1997 and Reich,

1991) have argued that regional technology development is controlled by international firms and markets, rather than local or national industrial policies. As a consequence of these approaches, national governments should not try to 'buck the market' by attempting to develop expertise in technological industries they can never hope to dominate. Instead, governments should allow international investment to build internationally-owned technology within their national territory, and then gain from the associated benefits of employment and secondary industries this investment brings.

Renewable energy technology is clearly of value to developing countries in terms of power supply, rural development or reduction of pollution. However, a key problem is to identify which renewable energy industries can be invited into developing countries without joint ventures or technological sharing, and which indigenous industries should be allowed to prosper without international competition. Evidence has suggested that some high-technology industries such as PV cannot survive when developed in isolation from global investment and the upgrading of product standards that results (see the example of Brazil, Chapter 3). However, this should not be taken to mean that developing countries should not attempt to develop new technological industries themselves, or that existing lower-technological industries in developing countries should be threatened by international investment.

A precedent for the impact of renewable energy development on technology ownership is the NFFO of the UK. This is commonly identified as a legislative measure that greatly increased investment in renewable energy technology in Britain. However, at the same time it also damaged the British wind turbine industry because it encouraged energy producers to use Danish turbines, which were at a more advanced stage of production. It is therefore important for governments to integrate investment in environmental or climate friendly technology with national industrial competitive strategies. Governments must decide on whether to prioritize investment in environmental technology from foreign sources, or the development of indigenous industries in the same technologies over a longer time period.

If the latter option is adopted, governments need to be confident that indigenous industries can develop a competitive advantage over foreign companies.

Using the CDM to subsidize foreign exports of climate technology, however, may have serious implications for local manufacturers of climate technology in non-Annex I countries. Providing incentives in the from of carbon credits for countries investing in the CDM is effectively to subsidize such investment, and therefore place investing companies at a competitive advantage over investors in countries who do not have access to these credits. The CDM may therefore place exporting countries at an unfair advantage over producers of renewable energy in developing countries, and therefore lead to the long-term erosion of the market share held by developing country companies.

Furthermore, if the CDM is used in this way, the result may be a new form of the 'hot-air problem' previously discussed in Chapter 1. The 'hot-air problem' usually refers to the problem in emissions trading when industrialized countries such as the USA may buy up the emissions quotas of countries that have undergone recent economic decline even though these countries are no longer producing carbon emissions on the same basis as the 1990 baselines indicate. For the CDM, a situation similar to 'hot-air' may arise when Annex I investors are allowed to claim credit for increasing sales of renewable energy technology when in fact all that has occurred is a shift in market share from developing country producers to investors from Annex I countries. The CDM may therefore be seen by developing countries to be a threat to local producers of climate technology, and also as another way to undermine the ambitions of the Climate Change Convention itself.

Flexible mechanisms for climate technology transfer

A main aim of this book has been to identify ways to enact international agreements on environment by harnessing business investment at the level of the firm. New flexible mechanisms of climate change mitigation, such as JI and the CDM can only be enacted by companies

> **Box 11.1: Flexible mechanisms for climate technology transfer**
>
> 1. Allow emissions reduction crediting activities undertaken by individual companies as well as by countries
>   – provides incentive for investment in carbon abatement, even without risking intellectual property (technology 'relocation', or vertical transfer).
>
> 2. Allow different levels of crediting for varying kinds of investment
>   – provides incentives for technology transfer projects of the greatest value to host countries or for climate change mitigation. Examples of preferred projects could include those assisting sustainable industrialization and capacity building rather than those that 'cherrypick' low-cost, low-risk ventures, or provide carbon sinks without transferring industrial technology.
>
> 3. Allow crediting for EST research and development at the national level, and disseminate technology through an EST bank or clearinghouse
>   – provides incentive for governments to invest in EST research, plus a store of publicly-owned technology which may then be shared or relocated by public or private bodies (horizontal transfer); also allows private companies to be credited for depositing technology, thus lessening the risk to IPR. The clearinghouse may also develop 'climate saving technology units' as a way to quantify the value of each new technological application.
>
> 4. Allow crediting for actions that build horizontal technology transfer
>   – provides incentives for companies or organizations which undertake more costly, long-term education and inculcation of technology use among new communities.
>
> 5. Create voluntary qualitative targets of EST research and development in non-Annex I countries
>   – allows integration of carbon abatement with economic growth (reversing the image that reducing greenhouse gases implies decreasing GDP. Also forms the basis of future South–South technology relocation/transfer.
>
> *Sources: the author, adapted from Chin, 1997; Chung, 1997*

acting on behalf of specific countries. It is therefore important to identify ways to integrate individual company actions into structures that contribute to national emissions targets. In effect, this is a redefinition of public–private responsibilities.

Box 11.1 lists five suggestions for implementing technology transfer through the combined activities of national public policies and private-sector investment. In effect, these make a new set of 'flexible' mechanisms for technology transfer that may be used in conjunction with the

CDM to reward companies or countries that undertake such actions by crediting them against QELROs.

The first strategy indicates the need to devise ways to integrate private investment with national emissions reductions targets. This requires careful monitoring of the impacts of each project on greenhouse gas emissions. The second suggestion is to reward different levels of crediting for actions with varying technology value. The third proposal is to credit public or private bodies which undertake research and development of climate technology and then place the technology in a public clearinghouse or 'technology bank'. Fourthly, investors should be rewarded for actions that undertake or build institutional capacity for horizontal technology. Fifthly, qualitative targets for technology development and transfer can be established for non-Annex I countries as a way to encourage South–South technology transfer.

These proposals may accelerate private international investment in technology transfer. However, two additional actions are required for these to take place. Firstly, there is a need for a climate technology index, to indicate the different values of each technology for mitigating climate change. This index could provide a valuation of each technology used or credited to a clearinghouse. Secondly, there is a need for the political apparatus to be developed in order to allow climate change mitigation to be undertaken at the level of private companies rather than simply Annex I countries. Firms may be able to collect credits and then trade these with countries where they are based.

In addition, many requirements to integrate private-sector actions into national emissions reductions targets depend on other debates concerning JI/AIJ and the CDM. In particular, it is not yet clear which baselines and scientific monitoring methods will be used for measuring the impacts of projects. Furthermore, the nature, composition and power of the CDM Executive Body have yet to be agreed. As noted in Chapters 1 and 2, much debate about the potential use of the CDM is for forestry-related projects rather than technology transfer. The ultimate integration of private investment into national emissions targets will depend on the resolution of these outstanding debates.

In the meantime, however, national governments and inter- and non-governmental organizations can increase the use of foreign investment in climate change mitigation by increasing the ease with which technology can be vertically and horizontally transferred. Evidence from the UK and the United States indicates that investment regulations such as the NFFO and PURPA have tended to increase investment in renewable energy (despite ownership implications). In South-east Asia, the SPP scheme in Thailand, PSKSK in Indonesia and the 'Pole-Vaulting' scheme of the Philippines may have the same effects. Governments may therefore empower private-sector investment for international environmental policy by strengthening and increasing these regulations, and undertaking associated actions such as decreasing subsidies to fossil fuels. Further investment capacity building by specialist organizations such as Preferred Energy Investments will also empower international investment regardless of the status of the climate change agreements.

Institutional finance and foreign investment

Direct investment

Institutional finance for renewable energy and climate-related technology investment is complex and evolving, with a number of trends occurring at the same time. As noted in Chapter 10, the old development aid approach of subsidizing new technology is now being replaced by the sustainable finance of specific projects. Cost recovery is now a major feature of technology development and transfer (see section 10.4). Furthermore, there are major changes occurring in the type of finance offered, and in the type of projects being supported.

The largest institutional lenders such as the World Bank are still, however, generally associated with large renewable energy projects such as dam construction and geothermal infrastructure. There is a need to increase institutional support to decentralized renewable energy development, but justifying some rural electrification schemes on financial returns can be difficult. Considering the macroeconomic advantages of rural electrification, reducing rural–urban migration, and increasing education and healthcare, make renewable energy costs more attractive.

Yet despite the relative advantages of different renewable energy finance schemes, the problem remains of whether technology projects are the most financially attractive for climate-related investment. This book has argued throughout that new mechanisms have to be found to provide an incentive for private investment in climate-related projects. In general terms, however, most investment for climate change mitigation to date has been in forestry-related projects because these offer better rates of return than technology development and transfer. Technology-related projects may be less competitive because of costs of technology development, joint ventures and the lack of an additional profit stream such as from forestry. If the CDM is intentionally organized to supply technology projects rather than forestry, it may find funds leaving it in favour of other existing climate-related funds that do focus on forestry, such as the World Bank Global Carbon Initiative (GCI). This may result in the failure of the CDM to attract sufficient finance in order to make technology transfer a priority of international investment, despite the fact that the CDM text in the Kyoto Protocol makes no mention of the word 'sinks'.

A precedent for this situation may be the Tropical Forestry Action Plan (TFAP), introduced by the FAO in 1985. The TFAP aimed to reduce tropical deforestation by integrating forest protection into overall development policy by identifying and listing various forestry-related projects in order to accelerate the financing of forest protection or conservation-linked activities such as ecotourism. Unfortunately, after a few years it became clear that finance had only been forthcoming for projects involving forest extraction rather than forest protection. Critics therefore claimed that the delineation of different projects into different purposes had actually increased investment in destructive land uses. Similarly, for the CDM and GCI, the potential evolution of two funding mechanisms for technology transfer and forest sequestration projects at different rates of return may ultimately undermine the ability of the CDM to attract funds.

A further uncertainty is the role of the GEF in relation to climate technology transfer. The GEF was established in 1990 by the World Bank, UNEP and UNDP as a fund to help address global environmental problems in developing countries. However, as noted in Chapter 1, its funding since 1990 has diminished partly as a result of the growth in private-sector transfers.

In many ways, the CDM threatens the GEF by providing finance for climate-change related investment. Under the terms discussed above, it is now possible for the CDM to provide investment specifically for climate change, and the GEF to address other concerns such as biodiversity conservation. Alternatively, the GEF may address the horizontal or institutional aspects of technology transfer, while the CDM could provide incentives for vertical technology transfer.

Portfolio investment

It is still unclear whether the CDM will take the form of a direct project-based funding device, or the form of a mutual fund for portfolio investment. If the CDM adopts a portfolio structure investors could place different sized donations for a share of the general return on investments undertaken by the fund. In this eventuality, the Executive Body of the CDM would have the power to decide which projects to support. However, the Executive Body would still be influenced by the need to attract funds to the CDM by addressing those projects popular with investors.

Experience of some sustainable development mutual funds in Britain has suggested that portfolio funds are primarily driven by the projects that are available to be funded by equity, and also by retail consumer interests. The result is that the funds' activities are shaped by market circumstances rather than by what might be considered to be the most urgent environmental action. It is therefore possible that the CDM might find itself under pressure to invest in forestry sequestration projects because these coincide with Northern investors' wishes and the misplaced belief that reforestation must also lead automatically to biodiversity conservation (see Chapters 1 and 2). The CDM may therefore not accelerate industrial technology transfer directly if investment

Redefining international investment and technology transfer 227

in it is dominated by populist retail concerns in industrialized countries. In response, governments and state electricity utilities in developing countries would aid the financial support for renewable energy or climate friendly projects by accelerating the equitization of renewable energy subsidiaries on stock exchanges.

Finally, there is also the question of whether the technology or the facilitation of investment requires support. As noted in Chapter 10, the trend in many renewable energy aid projects has been to support the financial capacity building for new renewable energy technologies rather than subsidizing the technology itself. The case studies have indicated the valuable role of intermediary organizations in the successful implementation of renewable energy technology transfer. Consequently, the successful institutional finance of foreign investment in renewable energy technology may be to support these organizations that let international investment succeed.

Conclusion and policy recommendations

This final section summarizes the arguments of the chapter, and draws lessons for practical policy options concerning the harnessing of foreign investment for climate change mitigation.

- Firstly, it is clear from the case studies that foreign investment and vertical transfer of technology alone are insufficient for effective technology transfer. There is much discussion of the need to harness international investment flows for environmental policy objectives. However, evidence suggests that the ability to direct private investment into a chosen sector (such as renewable energy) depends on the government policies that encourage investment, the accompanying regulatory environment and the facilitating organizations that reduce the transaction costs of investment, and also successfully embed new technologies into local economies.
- However, it is also clear that the approach adopted towards technology transfer in the UNFCCC negotiations should be amended to include vertical transfer (or relocation) of technology by multinational

producers which do not have to share this manufacturing knowledge with local users. This is consistent with new approaches to technology development based on the principles that local companies cannot compete in global high-cost markets, and that developing countries have much to gain from the associated benefits of international investment. But this does not mean an uncritical acceptance of free market forces. As Howells and Michie (1997:30) wrote, 'Globalization of technology does not imply the need for the abolition of national or regional policies, or an attempt to create a protectionist barrier around an economy's technology base; rather it requires sensitive policies that seek to engage the major economic base of the nation or region with both indigenous and foreign technological capabilities.'

- Enhancing technology transfer for climate change mitigation may therefore depend on a dual process of vertical and horizontal integration, in which foreign-owned technology is encouraged via investment but then is embedded into local economies via training and financial management in order to ensure its commercial success. It is, however, unclear where the responsibilities for horizontal embedding lie. In South-east Asia, the case studies suggest that specialized aid organizations such as CASE, SELF and Winrock International have succeeded in achieving local technology transfer. But few international investors have so far succeeded in embedding technology locally without the help of such organizations.
- Increasing international investment in renewable energy and other climate technology may, however, threaten the competitive standing of indigenous industries. There may be a trade-off between national industrial policy, seeking to increase international competitiveness through technological innovation, and international environmental policy that may support the immediate extension of international environmentally sound technology. The potential negative impacts of increased international investment are that international technologies such as PV turbines may replace domestic skills in biomass generation or potential competitors in PV and lead to a technological path dependency that benefits exporting countries. In

some cases, such as the so-called 'zero emission village', the dependency on imported PV technology may be more difficult to integrate with local users than existing technologies from local producers.
- International mechanisms for accelerating investment in technology such as the CDM may add to this potential conflict between international investors and local industries for two reasons. Firstly, they may effectively subsidize technology exports from Annex I countries without an equal benefit to non-Annex I countries. Secondly, they may also create a new investment-related 'hot air problem' in which companies from Annex I may be able to claim climate mitigating investments when in fact all that has been achieved is an increase in sales. These problems may be solved by encouraging some protection for developing world climate technology industries.
- International investment for climate technology transfer needs therefore to be considered as part of national industrial policy, with implications for the competitive standing of indigenous industries, and the trading terms with countries who face subsidies from undertaking such investment. The key requirement is to enhance international investment in climate-related technology, but also to ensure that this investment does not place local industries at threat, lessen local development by creating exploitative markets, or increase developing world technology dependency on Northern imports. For renewable energy, this may mean making broad distinctions between imported high technology such as PV, wind turbines and certain types of hydro and biomass generators; and locally produced technologies such as biogas generators. However, this should not be taken to mean that some developing countries cannot develop these higher forms of technology if they wish. Furthermore, evidence also suggests that technologies developed in the South are better suited for Southern users.
- Institutional finance for climate technology transfer is still evolving. However, there is a need to ensure that funding from large agencies is directed at small decentralized renewable energy rather than just centralized large hydro schemes. There is also a need to ensure that

new trends in using mutual funds for sustainable development projects do not become misdirected because of a lack of practical investment opportunities in developing countries, and a misplaced attention to populist environmental notions. Education of consumers by investors, and an increase in the equitization of technology concerns on stock exchanges is therefore necessary by developing world governments. Similarly, the GEF may be used alongside flexible mechanisms either for specifically non-climate related projects, or to assist in horizontal technology transfer.

- The future of climate technology-related funding arrangements is also uncertain if they are launched alongside similar funding bodies that focus specifically on forestry for carbon sequestration projects. As discussed in Chapter 2, the debate on the relationship of forestry and carbon sequestration is characterized by simplistic associations of reforestation and environmental impacts on both carbon sequestration, watershed control and biodiversity. International bodies need to work harder to ensure that JI/AIJ projects seeking reforestation for these reasons really do undertake the sufficient controls on tree species diversity and local implementation that are needed in order to ensure that these projects achieve the benefits they claim. Yet in addition, the generally better rate of return on forestry-related CDM or JI/AIJ projects implies that a specific technology-related fund may not attract sufficient funds if funds specializing in forestry, such as the World Bank GCI, exist alongside them. As a result, it may be necessary to require Annex I countries to undertake a specific proportion of carbon offset projects within technology fields – although these may be subject to monitoring for real impacts too, as discussed above.
- Ultimately, it is clear that foreign investment has great potential for enhancing the international transfer of environmentally sound technology. However, it must be harnessed in a way that reduces risk for investors, yet also channels investment into specific niches that are approved in a manner that consults with the public rather than simply reflects the least-cost options or the agendas of investors alone. There is also a necessity to assess the impacts of international investment on

local competitiveness and industrial policy. It is important to ensure that technology-related investment is seen to be valuable and profitable alongside forestry-related projects that also have potential negative impacts. The most immediate topics for clarification are which government policies may accelerate international investment in climate technologies quickest (possibly including laws such as the NFFO, SPP and PSKSK); and which agencies or investors should pay for the horizontal integration of imported technology.

Chapter 12

Enhancing public–private synergy in climate change policy

Introduction

This final chapter concludes the book by looking at the implications of the study for the enhancement and governance of foreign investment for climate change mitigation. As noted throughout the book, effective privatization and the harnessing of foreign investment flows between North and South does not simply depend on increasing private participation but also on establishing adequate incentives and regulations to ensure private investment supports public-sector objectives. Similarly, effective public–private synergy in climate change mitigation may also imply allowing vertical integration of international firms to increase economic efficiency with associated loss of competitiveness for other firms, and potentially a reduction in technological innovation.

This chapter considers these points by addressing the questions:

- What are the implications for national technological and competitive policies of harnessing foreign investment for climate change mitigation?
- How may environmentally sound technology be adopted locally if it is introduced via foreign investment and vertically integrated companies?
- What institutional factors may enhance international investment in renewable energy and other climate technology projects?
- How may privatization and liberalization of electricity supply industries in developing countries be managed to enhance environmental and developmental benefits?
- What are the different responsibilities of public and private sectors to ensure maximum gain for climate technology during industrialization and privatization, and what are the implications for funding?

Enhancing public–private synergy in climate change policy 233

The chapter is divided into three main sections. The first looks in detail at the implications of foreign investment and vertical integration for local technological development in developing countries, particularly renewable energy. The second section assesses the institutional factors that may enhance private international investment in climate change mitigation. The last section discusses the implications for creating effective business and governance partnerships for climate change mitigation that do not only increase returns for private investors, but also enhance public-sector objectives in both North and South.

Investment and national technology policy

The aim of this section is to assess which combinations of individual firm activity and national regulatory policy can maximize the development and adoption of climate technology. Two themes are addressed: first, the macro-scale management of industrial policy to enhance technology development, relating to foreign investment and potential international agreements such as the Multilateral Agreement on Investment, and second the smaller-scale implications for renewable energy and climate technology development.

Foreign investment and national competitiveness

In general terms, the impact of foreign investment on local development is managed to ensure that investment can achieve several policy objectives such as providing funds for government, spurring regional development and avoiding damage to the competitiveness of local industries. For investors, the risks are to avoid losing intellectual property rights through costly joint ventures and technology transfer schemes.

For climate change mitigation, environmental policy would be assisted by the rapid investment and adoption of climate technology regardless of the concerns of public and private sectors. Increasing investment and adoption of climate technology therefore brings risks to both local industries and to international investors.

As noted in Chapters 2 and 3, technology development is closely linked to industrialization. An established view is that national competitiveness in technology is related to the success of a nation's firms (Porter, 1990). However this view is being increasingly challenged on the grounds that some globally competitive high-technology markets cannot be mastered by indigenous technological development, and that nations should instead encourage international investment locally in order to gain the associated benefits such as employment and support for other industries (Reich, 1991). The degree to which individual industries might be suitable for local development may be controlled by the:

- mobility and immobility of assets (resources and factories);
- reproducibility and transferability of knowledge, information and expertise;
- homogenization and specialization in technological capability (Howells and Michie, 1997:225).

As a result of these factors, individual regions may develop specializations for technology production because of their ability to attract investment. Ironically, the process of 'globalization,' by which competitiveness and markets are becoming increasingly prevalent, may in fact lead to an increased specificity of local technological development in regions.

Furthermore, foreign investment may be divided according to whether companies seek markets, natural resources or production platforms on which to base further commercial activities (Esty and Gentry, 1997:160). Renewable energy and climate technology may in fact draw on all of these categories, but may also be seen to be unlike many traditional forms of resource exploitation such as mining or logging because renewable energy development is not generally destructive. As a result of this, local impacts of renewable energy development in developing countries through foreign investment may only affect competitiveness and extraction of profits rather than overt pollution.

Yet it is still unclear how foreign investment in renewable energy and climate technology may lead to increased adoption by local firms as either manufacturers or users of the product, or similarly how privatization may increase or decrease the access of poor people to electricity supplies. A common model of technology adoption – the theory of technological competence (Cantwell, 1994) – recognizes that foreign investment in technology may proceed when there is a time lag between demand for technology and its imitation by local producers. The model proposes that technological innovation is a cumulative process, and that innovation proceeds incrementally through small adjustments rather than large advances in a short amount of time (Freeman, 1992).

However, monopolization and vertical integration may not always lead to innovation or economic efficiency, and so in some cases the impacts of foreign investment on local development may slow down the adoption and innovation of climate technology (Johnstone, 1997). At times like this, the state may intervene to accelerate the adoption of technology through legislation that may either force companies to adopt existing technology, or develop new innovations (Wallace, 1995).

The role of the state in national development may often be overlooked. Evidence from East Asia has suggested that the most beneficial impacts of foreign investment have been in export-related industries, and under industrial policies that have also allowed some indigenous industries to develop under protection (Lall, 1996; Wade, 1990). This evidence has contradicted the common opinion that free market forces alone built industry in East Asia.

Consequently, a strong national industrial policy may be the most effective means to allow foreign investment to provide local benefits to economic development, but to minimize the impacts on indigenous industries. This suggests that demands for a Multilateral Agreement on Investment (MAI), which seeks to give international regulations on business and technological policy greater importance than national policy, may counter the attempts of developing countries to manage foreign investment for their own development. However, it seems clear that continued foreign investment may inevitably mean some loss of indigenous competitiveness and a growing regional specialization in export-oriented technology development.

> Overall, while international technology development within the firm is a consequence of the combination of the specific characteristics of technology in different firms and different locations, the distinctive competence of firms and countries is likely to be reinforced by it. Multinational enterprises that are able to integrate research activities internationally increase their technological competence, although this may also be associated with a greater diversity related technological activity. In contrast, countries are liable to become more technologically specialized, although governments may have some effect on the precise composition of such specialization. A central issue in determining which firms or countries benefit from the course they have become locked into is the variation in the growth of technological opportunities across sectors, and the nature of interrelatedness between different types of technological activity. (Cantwell, 1994:139)

Implications for climate technology

Foreign investment for renewable energy or climate technology development in developing countries may therefore lead to production becoming more regionally specialized but also increasingly internationally owned. Yet international investment in climate technology can only begin if it is seen to be profitable by investors and if government or other policies allow it. The case studies in Chapters 6 to 9 showed that many barriers to investment exist in countries such as Vietnam and Indonesia. Some intermediary organizations such as Preferred Energy Investments in the Philippines have been instrumental in lowering the transaction costs of market entry. Yet in general, foreign investment by itself is not sufficient to build either renewable energy development or technology transfer in rapidly developing countries.

An additional important question is the extent to which foreign investment can help produce appropriate technology for climate change mitigation and local development. As noted in Chapter 3, there is a tendency to assume that technological solutions require a North–South transfer of expertise. This belief may lead to high-technology

Enhancing public–private synergy in climate change policy 237

applications directed at global environmental concerns such as the PV-oriented 'zero emission village', rather than addressing the basic needs of farming communities in developing countries (see Chapters 10 and 11).

However, there is also a huge potential within developing countries for indigenous renewable energy development, particularly in biomass and occasionally in high-technology applications such as the PV industry in Indonesia. The question is how far such technologies can develop in the face of foreign investment and international competition before foreign investment creates a technological trajectory that has to be followed.

The available evidence is not supportive. In Chapter 3, a case study of PV development in Brazil revealed that local development of a globally competitive high-technology industry resulted in commercial failure because the manufacturer could not upgrade equipment to international standards. Similarly, evidence also suggests that simply subsidizing local development of technology outside market forces is unlikely to result in commercially sustainable industry, and that instead industries need to be developed immediately in conjunction with local cost recovery schemes (see Chapter 10). This evidence therefore does not bode well for, say, the Indonesian PV scheme that is currently being supported by ODA for the 'one million homes PV rural electrification scheme'. Ignoring this evidence, and persevering with local subsidized high-technology industries may eventually only slow climate change mitigation and rural electrification.

Furthermore, the Philippine 'Pole-Vaulting' scheme is also going against observed trends by firstly developing indigenous expertise in some renewable energy technologies, and secondly in attempting to increase technological progress in a rapid leap rather than incrementally. The availability of government funding for this grandiose scheme may also be under threat because of economic recession.

Another possible impact of foreign investment for renewable energy development is the 'dumping', or movement of outdated material from developed countries to developing countries where standards and expectations are lower. This is a common practice in many other industries but the impacts on local technological adoption are still not yet clear.

> The... global appropriation of technology in PV raises some questions on the linearity of innovation. Intramural and external knowledge sources are equally important for arriving at competitive levels of technology sophistication. Yet for relevant and growing market segments, not always the most advanced technology is required. (Grupp, 1997:200)

Consequently, the outlook for local renewable energy industries in developing countries looks bleak under international investment in high-technology applications, and the potential subsidizing of investment by the CDM. Regions are likely to become more specialized in producing renewable energy technology exports but the ownership of technology is likely to remain in foreign hands. Local governments can avoid negative impacts by taking actions that build incrementally on existing local expertise and resources such as in biomass, and integrating these into other sectors of government policy such as agriculture and forestry, and the treatment of agricultural and urban waste (see Box 9.3 concerning power from waste in the Philippines).

Furthermore, the role of energy efficiency measures must not be overlooked. In Thailand, this was shown to be successful in encouraging small electricity producers to use waste biomass to supply the national grid. In addition, effective energy efficiency measures may encourage portfolio investment. Governments may increase international investment in such enterprises by accelerating the equitization and launching of electricity companies and subsidiaries of the state utility on stock exchanges.

Electricity privatization in developing countries

The aim of this section is to identify how trends in foreign investment, technological development and climate change mitigation may be integrated with the liberalization and privatization of electricity supply industries in developing countries. The case studies in Chapters 6 to 9 revealed the nature of reforms and private-sector opportunities in four countries of South-east Asia. This section draws lessons from these case studies and identifies what steps can be taken to ensure that

Enhancing public–private synergy in climate change policy

market transition is integrated with environmental policy. Two themes are addressed: the market and governance structures emerging during privatization and liberalization, and the implications for renewable energy development.

Market and governance structures

Privatization and liberalization of electricity supply industries in developing countries is necessary in order to increase electricity supply, economic efficiency and to enhance investment in decentralized or renewable energy electricity generation. Some of the problems of unliberalized infrastructure development in developing countries are listed in Box 12.1. This box gives the results of a World Bank enquiry into economic efficiency in 1995.

Box 12.1: Critical constraints to private participation in Asia–Pacific infrastructure projects

1. Host countries have different perceptions of risks than investors – this causes delays in negotiations.
2. There is a lack of clarity about government objectives.
3. There is a lack of a credible and stable legal and regulatory framework.
4. Project risks are bundled together so that failure in one section may produce a failure in the whole project.
5. Domestic capital markets need to become more significant sources of finance in order to increase credibility and sources of funds.
6. There is a need for more sources of long-term debt.
7. There is a need to reduce transaction costs for government and investors, particularly through open competition.

Source: World Bank (1995), in Berg (1997:89)

The proposals in Box 12.1 indicate the need to increase private-sector freedom and sources of funding. Various forms of privatization have been applied, concerning different roles for the combination of market forces with responsible authorities in order to maximize public and private objectives (Berg, 1997). The initial, private-sector oriented approach may be divided into those that seek to increase efficiency through competition and reduce the power of monopolistic state utilities that decrease efficiency and innovation, and those that stress the need for reregulation rather than simply deregulation in order to ensure that new market structures do not replicate economic or social inefficiencies.

Evidence from the case studies reveals that the principle of free market competition does not yet apply to electricity privatization in Southeast Asia. Developed-world experience in electricity liberalization and privatization may not be a good model for developing countries because of the lack of well-developed markets and the regulatory structure for infrastructure projects. Furthermore, the early stage of development of many electricity sectors in Asia implies that most private-sector participation is to finance new construction of infrastructure rather than provide competition within existing bureaucracies or state-operated enterprises. Many BOT schemes also imply an eventual return of infrastructure to state ownership, and so the process of private-sector participation need not be linked to liberalization and decentralization.

Evidence also suggests that some state electric utilities in Asia are resistant to attempts to liberalize them. In Thailand, for example, the utility Egat has resisted initial government plans to break it into different sections on the grounds that some parts of electricity supply – such as the national grid – are natural monopolies and therefore need the continued presence of a centralized authority. The attempt to privatize and liberalize electricity supply industries in countries may therefore be limited to the general process of democratization and the ability to reduce the power of bureaucracies within the system of political reform. The existence of so-called natural monopolies in electricity supply industries should not be taken as a reason for avoiding reform to the bureaucracies that control them.

Liberalization and privatization can enhance private investment by reducing the transaction costs of market entry and operation. But the implications of reduced costs for rural electrification are unclear. One usual rationale for privatization is to end inefficient cross subsidization and allow clear project finance based on a PPA. But there may still be a need to provide funds for social or environmental objectives such as rural electrification, which might be avoided, if privatization is accompanied by strict cost accountability. Rural electrification may therefore become a stranded cost in the process of privatization unless there are provisions to continue some degree of cross subsidization. This may also mean that privatization reduces access for poor people to electricity unless some these measures are taken.

Effective privatization for public-sector objectives may therefore depend on the establishment of regulatory systems and public policy decisions before the process of privatization is started (Rosenzweig and Voll, 1997). The development of these structures varies in Asia. In the Philippines, for example, the local electric cooperatives and specialist agencies such as Preferred Energy Investments help establish local representation in private-sector activities. In Vietnam such political apparatus is poorly developed.

As a consequence, it may be necessary to implement full reforms such as the creation of an independent regulator or unbundling of new business units from state monopolies before privatization as a way to ensure that changes to electricity supply industries achieve public-sector priorities in both economic and developmental areas. These reforms may include, for example, the creation and equitization of subsidiaries for specific purposes such as renewable energy or rural electrification, and the establishment of fiscal regimes to ensure assistance to these projects based on the macroeconomic costs of not addressing rural electricity supply.

> To privatize the generation business without fundamentally strengthening the regulatory bodies and the transmission and distribution segment of the sector involves substantial risks for both private and public interests. If developing countries wish to attract substantial and continuous inflows of private capital, they must avoid the dangers of having to change the economic rules of the game at subsequent stages of restructuring. Otherwise, the availability of foreign capital will drastically decrease and its price will move up inexorably. (Bruggink, 1997:87)

Implications for climate change mitigation

The environmental implications of electricity privatization depend in part on which environmental objectives are sought. For developing countries, it is not always clear if investment and environmental policies seek to reduce greenhouse gas emissions, reduce energy wastage or encourage sustainable development in general – which may include decentralized rural electrification as a way to address basic needs.

The case studies have shown that some legislative reforms similar to Britain's NFFO exist in the form of Thailand's SPP laws and the PSKSK in Indonesia. However, the development of renewable energy in developing countries is clearly of secondary importance to increasing energy supply. Also, the options of replacing coal and oil use with natural gas, and then saving energy through DSM are also important climate change mitigating activities, although neither have been discussed in detail in this book.

Nevertheless, it is possible to enhance renewable energy development and rural electrification by integrating these with agriculture, forestry and waste management policies, as illustrated in Thailand and the Philippines. Renewable energy development may therefore be placed on the back of other – possibly more urgent – policies of increasing small private producers' participation in grid-supplied electricity generation, or dealing with the growing volumes of agricultural and urban waste.

There is also a clear need to increase renewable energy development through FDI and vertical technology transfer. The usual barriers to foreign investment in renewable energy markets are the high transaction costs of market entry and operation. However, the increased adoption of small, decentralized electricity generation systems effectively transforms the action of foreign investment from wholesale to governments, to retail to end consumers. This new trend may allow companies to bypass the state apparatus, especially if assisted by intermediary development organizations.

Furthermore, government agencies may themselves act as facilitators of new energy by adopting technologies and disseminating them throughout their network of buildings. The Telephone Organization

> **Box 12.2: Lessons for electricity privatization in developing countries**
>
> 1. Do not assume developed-world experience is a useful guide. Developing countries may have similar problems but need different solutions, particularly concerning the lack of national regulatory structures and competition.
>
> 2. Place more emphasis on transmission and distribution than on generation. The major challenge for many developing countries is to increase electricity supply to rural areas, rather than increase the variety of supply to the existing grid.
>
> 3. Isolate public policies from utility operation. Do not expect companies to act ethically in line with public policy objectives unless there are regulations and incentives to ensure that they do. Similarly, if public policy fails, blame the policymakers not the companies.
>
> Issues to be addressed by regulators:
> 1. Market structure reform and regulation of network industries
> 2. Financial analysis for utility regulation
> 3. Principles and application of incentive regulation
> 4. Non-price aspects of utility regulation
> 5. Managing the introduction of competition in and for the market
> 6. Rate structure
> 7. Managing the regulatory process for efficiency, transparency, credibility and legitimacy
>
> *Sources: Bruggink, (1997); Berg (1997)*

of Thailand, for example, has adopted PV technology for remote message relay centres, thus demonstrating the practicality of the product.

Privatization and liberalization of electricity supply industries in developing countries may therefore accelerate renewable energy development if governments are willing to create incentives and structures to focus new expenditure on renewable energy technology. However, evidence suggests that privatization and liberalization themselves are difficult to enforce in some developing countries because of opposition from state utilities such as Thailand's Egat, a lack of political will within governments, plus associated problems of corruption and the slowness of existing government bureaucracies.

However, progress may be made if the initial public policy reforms are in place before privatization begins, and if new renewable energy development is also linked to cost recovery. It may be necessary to adopt and strengthen legislation similar to the SPP and PSKSK of Thailand and Indonesia, and to justify rural electrification expenditures through their avoided macroeconomic costs. Box 12.2 lists three lessons of electricity privatization identified by two analysts in the context of developing countries, and some associated requirements for regulators.

Building public–private synergy in climate change mitigation

This final section considers the general implications of the book for creating effective business institutions for enacting inter-state agreements at the level of the firm. It is clear that harnessing foreign investment for climate change mitigation can bring many advantages of international technology transfer. However, foreign investment has to be channelled in the right direction by national and local regulations and incentives in order to maximize impacts. The aim of this section is to identify those business and governance structures that may direct foreign investment in this way.

Reducing transaction costs

Classically, debates about environmental regulation and business involvement have pointed to either market failure or government failure as the spur to increase or decrease the role of market forces. Seeking public–private synergy is an admission that both markets and governments have failures, and that both have to be addressed in consultation with each other.

Perhaps the most obvious market-related conclusion of this book is that foreign investment by itself is insufficient to enhance climate change mitigation or the adoption of renewable energy technology. Renewable energy development is clearly not simply an exercise in market penetration, and the development of sustainable businesses in renewable energy needs long-term embedding to increase adoption, and careful financial management to ensure commercial success.

Enhancing public–private synergy in climate change policy 245

Yet it is also clear that without foreign investment and privatization in developing countries, many opportunities for extending new technology and accelerating infrastructure development in order to avoid long-term commitment to fossil fuel sources of power would not exist. There is a need to reach a middle ground in which investment is accelerated through reducing transaction costs of market entry, yet technology is also embedded with local users.

The usual approach to reducing transaction costs is via vertical integration of firms. However, evidence from the case studies shows that other institutional forms can reduce transaction costs by assisting investors with local market research, training in new technology and building new financial management systems (see examples of CASE, Winrock International, Preferred Energy Investments, etc.). Indeed, in the UK and United States, the NFFO and PURPA laws achieved an influx of private investment that in time led to an overall decrease on costs for affected renewable energy technologies.

These institutional structures and incentives indicate that it is possible to reduce transaction costs by activities outside the firm, and through a process of horizontal integration between foreign investors and these intermediary organizations. The definition of business institutions for climate change policy in developing countries may therefore include complex alliances between vertically integrated firms and other support organizations. Without these alliances, both foreign investment and firms would be less effective. Furthermore, the alliances bring local benefits to host countries by integrating investment with local policy objectives such as technology transfer, rural electrification and waste disposal. International technology transfer may therefore in future also include establishing such new organizational structures.

The new institutional forms imply a variety of associated political and economic decisions. Enhancing vertically integrated investment implies a loss of competitiveness of some local industries. There is a need to increase the commercial structures of rural electrification: tariffs may have to be higher and include better accountability of financial management. Cross subsidization of renewable energy or rural electrification may still need to continue.

The important questions to be asked by national governments during the increase of foreign investment in renewable energy concern which technologies should be developed nationally or internationally, or which regulatory incentives and barriers should be imposed on investment. Eventually, some vertical integration may be reversed by governments if it is seen to be damaging. Yet the experience of Vietnam, where some investors such as Proctor and Gamble and Coca Cola are attempting to disband local joint ventures, suggests that such legislation may have to be enforced throughout the period of investment rather than simply in advance.

Responsibilities for public and private sectors

This book has argued that increasing technology transfer for climate change mitigation requires greater vertical integration and vertical transfer of technology by firms. However, both of these processes also require governance by national governments. Governance is needed in order to ensure that more investment is attracted to climate technology projects; that technology is embedded locally; and that local industries can avoid potential negative impacts of foreign investment on competitiveness or dependency on inappropriate technology.

This argument supports the general view that public–private synergy in public or environmental policy depends on perfecting the institutional structures within which private-sector investors operate, rather than believing in the power of markets forces alone to achieve public-sector policy objectives (Williamson, 1985). However, institutions can be interpreted in a variety of ways.

> The *institutional environment* is the set of fundamental political, social and legal ground rules that establish the basis for production, exchange and distribution. Rules governing elections, property rights and the right of contract are examples... An *institutional arrangement* is an arrangement between economic units that governs the way in which these units can cooperate and/or compete. It... [can] provide a structure within which its members can cooperate... or [it can] provide a mechanism that can effect a change in laws or property rights. (Davis and North, 19971:6–7, in Aoki *et al*, 1990:9)

In the context of climate change mitigation it is still unclear which institutional arrangements are possible or indeed desirable concerning foreign investment in developing countries. At present the trend of much policy is towards privatization. But as discussed above, the term is often used without appreciating the differences between countries and the political and economic context within which it is carried out. In essence, privatization can transfer technology and environmental standards and improve economic efficiency. But the implications are often that it leads to an increasing international ownership of technology and industry, and also an interpretation of environmental policy that suits companies rather than end users of technology.

> The rhetoric of the competition prescription has spread an ideological veneer over government practice, but the prescription itself speaks very little to the very real problem of making contracting work. (Kettl, 1993:199)

The political pitfalls of increased international investment in renewable energy via subsidies from the CDM include an erosion of local industries in developing countries; the introduction of high-technology energy sources that are poorly suited to local basic needs; and a failure to integrate international investment with local agricultural and forestry development. Furthermore, the CDM may also cause the problem of so-called 'investment hot air', in which Annex I countries may claim credit for mitigating climate change when in fact all they have done is to take market share from companies in developing countries (see Chapter 11).

It is therefore important that the Executive Body of the CDM, plus climate change negotiators and national governments, are aware of these pitfalls. International investment in renewable energy must be integrated with local resources and indigenous competitive industries, in order to produce appropriate technology and local development in developing countries.

A number of practical options are available. Official aid or CDM finance can go to reducing the transaction costs of marketing existing technology or the development of renewable energy by indigenous industries that are based on local resources and have competitive standing. Such industries might include local biomass facilities that have greater ability to be adapted by similar developing countries.

Governments in developing countries may also use their existing networks to adopt and demonstrate the use of new renewable energy technologies, such as the case of the Telephone Organization of Thailand and PV. However, such action may place further demands on an already shrinking state bureaucracy, in which case aid money could fund the adoption of technology for use in government networks.

There is also a general acknowledgement that investment in renewable energy may become more internationally owned and regionally specialized, and so damage local industries. Governments therefore need to adopt and strengthen new legislation similar to the UK's NFFO as a way to accelerate investment and use of renewable energy, although international domination of some technologies may be inevitable. In South-east Asia, the SPP and PSKSK legislation of Thailand and Indonesia may provide a format for these laws. However, there is also a need for great political will within government, and initial investments being justified on the basis of energy saving and rural development.

Coordinated action by governments has been useful:

> There is a direct relationship between the degree of government commitment at the top, and of the clarity of its objectives and the success a country has in attracting private investment in infrastructure. (World Bank, 1995, in Berg, 1997:97)

But it is important to note that all public policy decisions require an element of choice between priorities, for example, local indigenous competitiveness and international adherence to climate change policy. If rapid development of decentralized electricity leads to more investment and local development, then the fact the energy technology is foreign owned may not matter in the long run.

Finally, investors themselves have an important role. Multinational investment in renewable energy presents an alternative form of resource exploitation to other investments such as mining or logging. The long-term success of technology transfer is also assured by commercial embedding for cost recovery. Investors may therefore perform the crucial role of developing and marketing technology in a way that is more economically sustainable than state-operated institutes, and then work to embed technology in cooperation with development organizations or NGOs seeking to advance environmental policy or local development. Ironically, this was illustrated in 1998 by a joint venture between Shell and Greenpeace to distribute PV in the Netherlands. This alliance of classic enemies in the growing market of renewable energy suggests that apparently deep-seated impasses may be overcome.

Conclusion and policy recommendations

> The modern nation state, some 200 years old, is no longer what it once was: vanishing is a nationalism founded upon the practical necessities of economic independence within borders and security against foreigners outside. (Reich, 1991:315)

> The winds of creative destruction are increasingly foreign. (Howells and Michie, 1997:222)

This chapter has supplemented conclusions in Chapters 10 and 11 in regard to the policy implications of involving business investment in climate change policy. The chapter assessed the impacts of foreign investment on national competitiveness and technological development; the governance of electricity privatization in developing countries; and the options for creating public–private synergy in climate change policy. The following is a list of some of the chapter's key conclusions.

- Integrating foreign investment and climate change mitigation presents many potential options for enhancing technology transfer and environmental standards. But there are also additional impacts on other areas of public policy that need to be acknowledged.
- Most obviously, harnessing foreign investment for climate change mitigation involves a loss of competitiveness for indigenous industries. There is also a risk of encouraging dependency on technological

trajectories that are inappropriate for some users of electricity in developing countries, particularly in poor rural locations. It is important not to let the privatization and liberalization of state monopolies in developing countries become a new dependency on internationally vertically integrated companies with alternative problems for local development. As a result, sometimes it may be more effective for climate change mitigation and local development to enhance investment in existing renewable industries in developing countries.

- Similarly, simply encouraging foreign investment itself is not enough to build renewable energy investment. There is a need for national and local regulations, incentives and organizations to accelerate investment in particular industries and then embed new technology in local economies and day-to-day practices. There is consequently a need for flexible national industrial policies within developing countries – something that the MAI could have threatened.
- Technology innovation and adoption in regions is likely to be incremental and linked to those industries that can compete in global markets. Unfortunately, this suggests that ambitious rapid development plans using indigenous high-technology industries in developing countries (such as the Philippines' 'Pole-Vaulting' renewable energy scheme) may fail unless it can harness international investment in high-technology applications.
- Privatization of electricity supply industries in developing countries presents a variety of options for increasing investment in decentralized or renewable energy systems. However, privatization by itself without a guiding regulatory framework may result in an increased efficiency of existing grid-supplied power generation, and decreases access for poor people rather than a radical alteration of existing policies. In many developing countries conservatism and poor implementation of policies within bureaucracies, and a lack of market competition may make the overall aims of privatization and liberalization difficult to achieve.
- Consequently, technology transfer needs vertical integration and vertical forms of transfer (see Chapter 3). But vertical integration

and transfer need governance. Governance structures need to identify priorities of public policy such as local industrial development, adherence to global environmental policy or attraction of international investment because of associated benefits. However, many local forms of managing technology transfer resulting from vertically integrated companies can be achieved by the action of intermediary development organizations who reduce transaction costs of investment, and integrate new energy with local development objectives.

- The implications of harnessing foreign investment for climate change mitigation need to be discussed further within the climate change negotiations. At a time when policymakers are increasingly looking to private investment to fulfil public-sector objectives it is crucial that the possible implications are identified and managed. At best, foreign investment for climate change mitigation may transfer technology and enhance local development in developing countries. At worst, investment subsidized by the CDM could destroy the competitiveness of local industries, and achieve only 'investment hot air' – or the claim that investment has helped mitigate climate change, when in fact all investors have done is increase market share at the expense of manufacturers in developing countries.

Appendix 1

Article 12 of the Kyoto Protocol: the Clean Development Mechanism

1. A Clean Development Mechanism (CDM) is hereby defined.

2. The purpose of the CDM shall be to assist Parties not included in Annex I in achieving sustainable development and in contributing to the ultimate objective of the Convention, and to assist Parties included in Annex I in achieving compliance with their quantified limitation and reduction commitments under Article 3.

3. Under the CDM:

- Parties not included in Annex I will benefit from project activities resulting in certified emission reductions; and
- Parties included in Annex I may use the certified emissions reductions accruing from such project activities to contribute to compliance with part of their quantified emission limitation and reduction commitments under Article 3, as determined by the Conference of the Parties serving as the meeting of the Parties to this Protocol.

4. The CDM shall be subject to the authority and guidance of the Conference of the Parties serving as the meeting of the Parties to this Protocol and be supervised by an executive board of the CDM.

5. Emissions reductions resulting from each project activity shall be certified by operational entities to be designated by the Conference of the Parties serving as the meeting of the Parties to this Protocol, on the basis of:

- voluntary participation approved by each Party involved;
- real, measurable and long-term benefits related to the mitigation of climate change; and
- reductions in emissions that are additional to any that would occur in the absence of the certified project activity.

6. The CDM shall assist in arranging funding of certified project activities as necessary.

7. The Conference of the Parties serving as the meeting of the Parties to this Protocol shall, at its first session, elaborate modalities and procedures with the objective of ensuring transparency, efficiency and accountability through independent auditing and verification of project activities.

8. The Conference of the Parties serving as the meeting of the Parties to this Protocol shall ensure that a share of the proceeds from certified project activities is used to cover administrative expenses as well as to assist developing country Parties that are particularly vulnerable to the adverse effects of climate change to meet the costs of adaptation.

9. Participation under the CDM, including in activities mentioned in paragraph 3(a) above and in acquisition of certified emission reductions, may involve private and/or public entities, and is to be subject to whatever guidance may be provided by the executive board of the CDM.

10. Certified emission reductions obtained during the period from the year 2000 up to the beginning of the first commitment period can be used to assist in achieving compliance in the first commitment period.

Appendix 2

Brief summary of renewable energy technologies

Solar photovoltaic systems

The conversion of ultraviolet light into electricity. To maximize output, single solar cells are arranged in arrays and mounted in weather-proof modules. New developments here include the production of malleable sheets of PV circuitry, allowing end users to integrate PV coverage with individual buildings and the different angles of sunlight. Solar PV systems are among the most advanced versions of renewable energy technology.

Solar thermal systems

The harnessing of solar energy to produce hot water. Water is passed through pipes exposed to sunlight and then sent to a storage tank, usually for domestic use. The water flow may either be natural, resulting from gravity or convection within the circuit, or induced artificially by the use of a pump. Solar thermal systems are comparatively unsophisticated in relation to PV, and the most common problems associated with them relate to the water either leaking, freezing or depositing calcium carbonate inside the pipes. Solar thermal systems are also known as passive solar because they simply transfer heat rather than generate electricity.

Small hydroelectric power

The insertion of a small turbine into a river, dam or canal in order to operate a nearby generator. Electricity is then transmitted to local users via transformers and transmission lines. There are two forms of hydraulic turbines. Impulse turbines such as the Pelton type stand partially clear of water flow. Reaction turbines, such as the Francis or propeller turbine are completely immersed in the flow. Water turbines in small hydroelectric power may have efficiency of more than 90 per cent, although costs may rise in locations where the water head (or pressure) is low.

Biomass energy systems

The combustion of vegetative matter (commonly wood) to create heat or gas, which may then be used for generating electricity. This form of energy is counted

Brief summary of renewable energy technologies 255

as renewable because the biomass fuel may be regrown; the growth of plants sequesters carbon; and biomass has lower sulphur concentrations than fossil fuels. However, biomass has lower energy density than fossil fuels (for example, wood has 18 GJ/t compared with 30 GJ/t for coal), and so biomass use has to be close to the place of collection in order to minimize transport costs. Costs also rise according to the moisture context of biomass.

There are various types of biomass technology:

Wood-fired power plants/stoker boilers: these are the least sophisticated, and involve the direct combustion of biomass inside incinerators. Potential problems with these power plants are corrosion and stresses caused by moisture content and the accumulation of ash.

Fluidized-bed combustors: these aim to reduce SO_2 and NO_x emissions and increase the variety of fuels burned inside combustors by providing a platform inside the incinerator for fuels to stand, and also allowing side vents into which air may be pumped.

Biomass gasification: gasification is a particular form of combustion in which hydrocarbon gases are created through feeding biomass fuel over a bed of coals, and then piped to use in other applications. This has the advantage of reducing tar deposits within combustion units, and is generally of value for small-scale applications. Gasifiers can also be used with a variety of agricultural waste such as rice husks, charcoal, coconut shells, corn cobs and cassava stalks.

Anaeobic digestion/biogas digesters: biogas is a mixture of methane (50–70 per cent) and carbon dioxide, and may be produced under anaerobic conditions (ie in the absence of air or free oxygen). Biogas is created by mixing water with organic material such as cow dung, and then placed to ferment inside a digester. The resulting gas is used to make a combustible gas, and the remaining waste is valuable as a fertiliser. Certain types of digester have been identified according to the region where they are used:

- the floating drum (or Indian) digester is based in a cylindrical brick pit in the ground, in which a steel drum collects the gas as it is emitted.
- the fixed dome (or Chinese) digester, is again in an underground pit, but has a domed cover held in place by soil deposits.
- the flexible bag (or Taiwanese) digester uses a plastic cylindrical bag supported in a trench lined with masonry, concrete or mud.
- there are also other, more advanced, designs such as the upflow anaerobic sludge blanket (UASB) design, which are popular when integrated with urban or industrial waste including at sewage works, abattoirs or breweries. Landfill, or urban waste, may also produce methane, which can then be used for power generation.

Wind energy systems

The use of wind-driven propellers or rotors to generate electricity or pump water. Water pumping is by far the most common use of wind energy in developing countries, but this is being challenged gradually by the influx of new technology that allows the motion of the rotor to generate electricity through an attached turbine.

Hybrid systems

The combination of one or more forms of renewable energy with fossil fuel sources in order to increase reliability of supply, or reduce the GHG emissions of existing non-renewable power sources. One example is the co-firing of coal with biomass in large power stations in order to reduce GHG emissions.

'Large-scale' renewable energy technologies
Geothermal energy

The use of heat deep inside the earth to heat water for electricity generation. Wells are drilled in zones of known geothermal activity in order to pump water into a warming zone before harnessing it when it emerges at a higher temperature.

Large hydropower (dam) energy

The construction of dams in order to generate electricity by releasing water through turbines at high pressure. These have severe institutional barriers to development because of the need to resettle villages or flood forest or agricultural land, and the high costs of development. Dams have also been linked to increases in diseases like malaria and bilharzia.

Ocean energy

There are three general ways for utilizing oceans for renewable energy:

Tidal energy: the insertion of turbines in coastal zones to exploit tidal currents or wave energy. The main constraint here is the interference with coastal sea traffic and fishing.

Ocean thermal energy conversion (OTEC): the exploitation of thermal differentials between warm surface water and cold deep ocean water in open sea locations. The flow of water induced by pumping water may drive turbines. The constraints of OTEC include high costs and low efficiency. The theoretical maximum effi-

Brief summary of renewable energy technologies 257

ciency for a temperature differential of 20 °C is just 6.8 per cent, and 9.0 per cent for 27 °C. In reality, efficiencies are between 2.5 and 4 per cent.

Osmotic pressure of fresh and saltwater: The release of freshwater into saltwater would produce an increase in the physical volume of the water (if salinity is constant), resulting in a flow of water to power turbines. However, this third approach has been poorly developed and is generally considered expensive.

Sources: World Energy Council, 1994; Green, 1996:4–26

References

Adger, N.; Pettenella, D.; and Whitby, M. (eds) (1997) *Climate change mitigation and European land-use policies*, Wallingford: CAB International

Ahmed, K. (1994) *Renewable energy technologies: a review of the status and costs of selected technologies*, World Bank Technical Paper No.240, Energy Series, Washington D.C.: World Bank

Alchian, A. and Demsetz, H. (1972) 'Production, information costs and economic organisation', *American Economic Review* 62:5 777–795

Alston, L.; Eggertsson, T.; and North, D. (eds) (1996) *Empirical studies in institutional change*, Cambridge: Cambridge University Press

Amranand, P. (1993) 'Thailand's privatisation programme – progress to date', pp.3.1–3.4 in *Asian electricity: the growing commercialisation of power generation*, FT Conference, Singapore 25–26 May 1993, Wolverhampton: The Freelance Association

Andersen, R. (1997) *Challenges for sustainable energy sectors in developing countries: with case studies from Zambia, Zimbabwe, India and Thailand*, EED Report 1/1997, Lysaker, Norway: Fridtjof Nansen Institute

Baldwin, S.; Burke, S.; Dunkerley, J. and Komor P. (1992) 'Energy technologies for developing countries: US policies and programs for trade and investment', *Annual Review of Energy and the Environment* 17: 327–358

Barnett, A.; Bell, M. and Hoffman, K. (1982) *Rural energy and the Third World: a review of social science research and technology policy problems*, Oxford: Pergamon Press

Berg, S. (1997) 'Priorities in market reform: regulatory structure and performance', *Pacific and Asian Journal of Energy* 7:2 89–102

Bhagat, R. (1993) *Rural electrification and development*, New Delhi: Deep and Deep Publications

Birchall, J. (1997) 'BOT slot: build-operate-transfer projects are making progress, but slowly', *Vietnam Economic Times* 45, November 1997, pp.28–29

Blair, R. and Kaserman, D. (1983) *Law and economics of vertical integration and control*, New York: Academic Press

Brack, D. (1995) *International trade and the Montreal Protocol*, London: Earthscan/RIIA

Breeze, P. (1996) *Electricity in Southeast Asia*, London: FT Energy

Brown, K.; Adger, W. and Tuner, R. (1993) 'Global environmental change and mechanisms for North-South resource transfers', *Journal of International Development*, 5:6 571–589

Bruggink, J. (1997) 'Market metaphors and electricity sector restructuring: lessons for developing countries', *Pacific and Asian Journal of Energy* 7:2 81–88

Buckley, P. and Casson, M. (1991, 2nd ed) *The future of the multinational enterprise*, London: Macmillan

Bush, E. and Harvey, L. (1997) 'Joint implementation and the ultimate objective of the UNFCCC', *Global Environmental Change* 7:3 265–286

Buttel, F. and Taylor, P. (1994) 'Environmental sociology and global environmental change', pp228–255 in Redclift, M. and Benton, T. (eds) *Social theory and the global environment*, London: Routledge

Cairncross, F. (1995) *Green Inc.: a guide to business and the environment*, London: Earthscan
Cantwell, J. (1989) *Technological innovation and multinational corporations*, Oxford: Blackwell
Cantwell, J. (1994) 'The theory of technological competence and its application to international production', pp.37–72 in Cantwell, J. (ed) *Transnational corporations and innovatory activities*, UN Library on Transnational Corporations vol.17, London: Routledge
Casson, M. (ed) (1986) *Multinationals and world trade: vertical integration and the division of labour in world industries*, London: Allen and Unwin
Casson, M. (1987) *The firm and the market: studies in multinational enterprise and the scope of the firm*, Oxford: Blackwell
Caves, R. (1996, 2nd ed.) *Multinational enterprise and economic analysis*, Cambridge: Cambridge University Press
Charles, D. and Howells, J. (1992) *Technology transfer in Europe: public and private networks*, London: Belhaven
Chung, Rae Kwon (1998) 'The role of government in the transfer of environmentally sound technology', pp47–62 in Forsyth, Tim (ed) (1998) *Positive measures for technology transfer under the Climate Change Convention*, London: RIIA
Clarke, J. (1983) *Issues of governance and international investment and multinational enterprises*, Wye Paper, New York: Aspen Institute for Humanistic Studies
Clarke, T. and Pitelis, C. (eds) (1993) *The political economy of privatisation*, London: Routledge
Coase, R. (1937) 'The nature of the firm', *Economica* 4:16 (NS) 386–405
Coase, R. (1964) 'The regulated industries: discussion', *American Economic Review* 54: 194–197
Collamer, N. and Rose, A. "The changing role of transaction costs in the evolution of joint implementation", *International Environmental Affairs* 9:4 274–288
Corry, S. (1994) *'Harvest moonshine' taking you for ride: a critique of the 'rainforest harvest' – its theory and practice*, London: Survival International
Coughlin, C. (1983) 'The relationship between foreign ownership and technology transfer', *Journal of Comparative Economics*, 7: 400–414
Cullet, P. and Kameri-Mbote, P. (1998) 'Joint implementation and forestry projects: conceptual and operational fallacies', *International Affairs* 74:2 393–408
Davis, L. and North, D. (1971) *Institutional change and American economic growth*, Cambridge: Cambridge University Press
Dixon, C. (1998) *Thailand: uneven development and internationalisation*, London: Routledge
Dhiratatakinant, Kraiyudht (1989) *Privatisation: an analysis of the concept and its implementation in Thailand*, Bangkok: Thailand Development Research Institute
Djojodihardjo, H.; Mulyadi, R ; Dasuki, A.; and Djamin, M. (1997) 'The one-million homes photovoltaic rural electrification program in Indonesia', paper presented at the *Asia Pacific Initiative for Renewable Energy and Energy Efficiency Conference*, 14–16 October 1997, Jakarta
Dunning, J. (1988) *Multinationals, technology and competitiveness*, London: Unwin Hyman
Dunning, J. (ed) (1998) *Globalisation, trade and foreign direct investment*, Oxford: Pergamon
Dunning, J. and Narula, R. (eds) (1996) *Foreign direct investment and governments: catalysts for economic restructuring*, London: Routledge
Earl, G. (1998) 'Indonesia pushes energy projects despite crisis: .crisis can't dim region's energy hopes', *Australian Financial Review* 4th August 1998, page number unknown

Eden, S. (1996) 'The politics of packaging in the UK: business, government and self-regulation in environmental policy', *Environmental Politics* 5:4 632–653

Enos, J. and Park, W. (1988) *The adoption and diffusion of imported technology: the case of Korea*, London: Croom Helm

Esty, D. and Gentry, B. (1997) 'Foreign investment, globalisation and environment', pp141–172 in OECD, *Globalisation and environment: preliminary perspectives*, Paris: OECD

Fabris, A. and Servant, M. (1996) 'Argentina Dispersed Rural Population Electricity Supply Program', Secretaria de Energia, Buenos Aires, Argentina, 22 October 1996

Fairhead, J. and Leach. M. (1998) *Reframing deforestation: the global and the local, studies in West Africa*, London: Routledge

Foray, D. and Freeman, C. (eds) (1993) *Technology and the wealth of nations: the dynamics of constructed advantage*, London: Pinter

Freeman, C. (1987) *Technology policy and economic performance: lessons from Japan*, London: Pinter

French, H. (1998) *Investing in the future: harnessing private capital flows for environmentally sustainable development*, Worldwatch Paper 139, Worldwatch Institute: Washington D.C.

Forsyth, D. (1990) *Technology policy for small LDCs*, London: Macmillan

Forsyth, T. (1997) 'Environmental responsibility and business regulation: the case of sustainable tourism', *The Geographical Journal* 163:3 270–280

Forsyth, T. (ed) (1998) *Positive measures for technology transfer under the Climate Change Convention*, London: RIIA

FOE (Friends of the Earth) (1995) *A superficial attraction: the voluntary approach and sustainable development* London: FOE

Gamba, J.; Caplin, D. and Mulckhuyse, J. (1986) *Industrial energy rationalisation in developing countries*, Baltimore and London: John Hopkins University Press

Garrod, B. (1997) 'Business strategies, globalisation, and environment', pp269–314 in OECD, *Globalisation and environment: preliminary perspectives*, Paris: OECD

Ghozali, Y. (1993) 'Malaysia's growing electricity requirements – private-sector business opportunities', pp1.1–1.4 in *Asian electricity: the growing commercialisation of power generation*, FT Conference, Singapore 25–26 May 1993, Wolverhampton: The Freelance Association

Graham, E. (1978) 'Trans-Atlantic investment by multinational firms: a rivalistic phenomenon?' *Journal of Post-Keynesian Economics* 1:1

Gregory, J.; Silveira, S.; Derrick, A.; Cowley, P.; Allinson, C.; and Paish, O. (1997) *Financing renewable energy projects: a guide for development workers*, London: Intermediate Technology Publications, in association with the Stockholm Environment Institute

Goldberg, D. and Stilwell, M. (1997) *Twelve principles to guide joint implementation*, Center for International Environmental Law, Washington/Geneva, October 1997

Gouldson, A. and Murphy, J. (1998) *Regulatory realities: environmental policy and implementation in the UK*, London: Earthscan

Green, J. (1997) *Renewable energy systems in Southeast Asia*, Tulsa, Oklahoma: Pennwell

Grubb, M. (1995) *Renewable energy strategies for Europe, volume I: Foundations and context*, London: Earthscan/RIIA

Grubb, M., with Vigotti, R. (1997) *Renewable energy strategies in Europe, volume II: electricity systems and primary electricity sources*, London: Earthscan/RIIA

Grubb, M.; Koch, M.; Munson, A.; Sullivan, F.; and Thomson, K. (1993) *The Earth Summit agreements: a guide and assessment*, London: Earthscan and RIIA

Grupp, H. (1997) 'Technical change on a global market: competition in solar cell development', pp177–202 in Howells, J. and Michie, J. (eds) (1997) *Technology, innovation and competitiveness*, Cheltenham: Edward Elgar

Gupta, J. (1997) *The Climate Change Convention and developing countries: from conflict to consensus?*, London, Dordrecht, Boston: Kluwer

Ha-Duong, M.; Grubb, M.; and Hourcade, J. (1996) 'Optimal emission paths towards CO_2 stabilization and the cost of deferring abatement: the influence of inertia and uncertainty', Working paper, CIRED, Montrouge, France

Hajer, M. (1995) *The politics of environmental discourse: ecological modernisation and the policy process*, Oxford: Clarendon Press

ul-Haque, I. with Bell, M. (1998) *Trade, technology and international competitiveness*, Washington D.C.: World Bank

Harriss, J.; Hunter, J.; and Lewis, C. (eds) (1995) *The new institutional economics and Third World development*, London: Routledge

Harvie, C. (1997) 'Energy supply and demand and its contribution to economic development in a transition economy: the case of Vietnam', paper presented at the IAEE conference, San Francisco, November 1997

Harvey, L. and Bush, E. (1997) 'Joint implementation: a strategy for combating global warming?', *Environment* 39:8 14–20, 36–43

Head, C. (1994) 'Vietnam's changing hydro scheme', *International Water Power and Dam Construction* March 1994 pp.27–28

Heaton, G.; Banks, R. and Ditz, D. (1994) *Missing links: technology and environment implications in the industrialising world*, Washington D.C.: World Resource Institute

Heller, T. (1998) 'Joint Implementation, transactional costs and the political economy of climate change', pp99–116 in Forsyth, Tim (ed) (1998) *Positive measures for technology transfer under the Climate Change Convention*, London: RIIA

Hirsch, P. and Warren, C. (eds) (1998) *The politics of environment in Southeast Asia: resources and resistance*, London: Routledge

Howells, J. and Michie, J. (eds) (1997) *Technology, innovation and competitiveness*, Cheltenham: Edward Elgar

Hurst, C. and Barnett, A. (1990) *The energy dimension: a practical guide to energy in rural development programmes*, London: Intermediate Technology Publications

Hymer, S. (1976) *The international operations of national firms: a study of direct investment*, Cambridge, Mass.: MIT Press

IEA (International Energy Agency) (1994) *Demand side opportunities and perspectives in the Asia Pacific region, with emphasis on the gas and electricity sectors: proceedings of a conference at Seoul, Korea 4–5 September 1993*, Paris: OECD and IEA

IEA (International Energy Agency) (1995) *Energy, technology and developing countries: a study of four countries*, Paris: OECD and IEA

IEA (International Energy Agency) (1996a) *International energy technology collaboration, benefits and achievments*, Paris: OECD and IEA

IEA (International Energy Agency) (1996b) *Competition and new technology in the electric power sector*, Paris: OECD and IEA

IEA (International Energy Agency) (1997a) *Asia electricity survey*, Paris: OECD and IEA

IEA (International Energy Agency) (1997b) *Renewable energy in IEA countries, Volume 1: overview*, Paris: OECD and IEA

IEA (International Energy Agency) (1997c) *Key issues in developing renewables*, Paris: OECD and IEA

Intarapravich, D.; Lim. L.; Resanond, A.; Santisirisomboon, J.; Shrestha, A., and Decosta, I. (1995) *Electricity: meeting needs with least environmental impacts*, Thailand Environment Institute 1995 Annual Conference: Sharing environmental responsibility – towards partnership management, Bangkok: TEI

Ives, J. and Messerli, B. (1989) *The Himalayan dilemma: reconciling conservation and development*, London, New York, Tokyo: United Nations University

Janssen, H., Kiers, M. and Nijkamp, P. (1995) 'Private and public development strategies for sustainable tourism development of island economies, pp65–84 in Coccossiss, H. and Nijkamp, P. (eds) *Sustainable tourism development*, Aldershot: Avebury

Jepma, C. (1995) (ed) *The feasibility of joint implementation*, Dordrecht: Kluwer Academic Publishers

Johnstone, N. (1997) 'Globalisation, technology and environment', pp227–268 in OECD, *Globalisation and environment: preliminary perspectives*, Paris: OECD

Joskow, P. (1985) 'Vertical integration and long-term contracts: the case of coal burning electric generating plants', *Journal of Law, Economics and Organization* 1:1 33–80

Karekezi, S. (1992) 'Disseminating renewable energy technologies in sub-Saharan Africa', *Annual Review of Energy and the Environment* 19: 387–421

Kay, N. (1993) 'Markets, false hierarchies and the role of asset specificity', pp.242–265 in Pitelis, C. (ed) *Transaction costs, markets and hierarchies*, Oxford: Blackwell

Kennedy, E., and Williams, P. (1998) *Lessons learned and being learned: successful models for capacity building*, Briefing paper, Winrock International Renewable Energy Program: Arlington, Virginia

Kettl, D. (1993) *Sharing power: public governance and private markets*, Washington D.C.: The Brookings Institution

Kirtikara, K. (1997) 'Applied PV in developing countries: photovoltaic applications in Thailand: twenty years of planning and experience', paper presented at the Mekong River Basin PV Seminar, 19–22 March 1997, King Mongkut Institute of Technology, Thonburi, Bangkok

Khan, A. (1985) *Foreign investment and technology transfer: fiscal and non-fiscal aspects. Country profiles on fiscal and non-fiscal regimes affecting the two-way flow of investment and technology transfer between the developed countries and the Asia-Pacific region*, Singapore: Asian-Pacific Tax and Investment Research Centre

Kozloff, K. (1994) *Rethinking development assistance for renewable energy*, World Resources Institute: Washington D.C.

Kozloff, K. (1995a) 'Achieving India's potential for renewable electricity: policy lessons from the United States', pp7–26 in Ramana, P.V. and Kozloff, K. (eds) *Renewable energy development in India: analysis of US policy experience*, New Delhi: Tata Energy Research Institute

Kozloff, K. (1995b) 'Rethinking development assistance for renewable energy sources', *Environment* 37:9 6–32

Kozloff, K. (1988) 'Electricity sector reform in developing countries: implications for renewable energy', occasional paper published on the internet by REPP (www.repp.org)

Lall, S. (1996) *Learning from the Asian Tigers: studies in technology and industrial policy*, Basingstoke: Macmillan

Lan, P. and Young, S. (1996) 'Foreign direct investment and technology transfer: a case study of foreign direct investment in north-east China', *Transnational Corporations* 5:1 57–84

Leach, G. and Mearns, R. (1988) *Beyond the fuelwood crisis*, London: Earthscan

Lee, Yeong Heok (1994) *Vertical integration and technological innovation: a transaction cost approach*, New York: Garland

Lefevre, T. and Bui Duy Thanh (1996) 'Energy efficiency and conservation in Thailand: analysis of the 1980–1992 period', *Energy Studies Review* 8:2 155–166

Lefevre, T.; Pacudan, R.; and Todoc, J. (1997a) *Power in the Philippines: market prospects and investment opportunities*, London: FT Energy

Lefevre, T.; Pacudan, R.; Timilsina, G.; and Todoc, J. (1997b) *Power in Thailand: market prospects and investment opportunities*, London: FT Energy

Leonard-Barton, E. (1990) 'The intraorganisational environment: point-to-point versus diffusion', pp.43–62 in Williams, F. and Gibson, D. (eds) *Technology transfer: a communication perspective*, London: Sage

Loske, R. and Oberthür, S. (1994) 'Joint implementation under the climate change convention: opportunities and pitfalls', *International Environmental Affairs* 6: 45–58

Marquand, C.; McVeigh, J.; and Sehgal, S. (1998) 'Dissemination of renewable energy technologies in developing countries – lessons for the future', *International Journal of Ambient Energy* 19:1 3–33

Marsudi, D. (1997) 'Indonesia electric power development: role of the private sector and development of renewable energy', paper presented at the *Asia Pacific Initiative for Renewable Energy and Energy Efficiency Conference*, 14–16 October 1997, Jakarta

Martinot, E.; Sinton, J.; Haddad, B. (1997) 'International technology transfer for climate change mitigation and the cases of Russia and China', pp357–402 in *Annual Review of Energy and the Environment volume* 22, edited by Socolow, Robert H.; Anderson, Dennis, and Harte, John

Mashayekhi, A. (1992) *Review of electricity tariffs in developing countries during the 1980s*, World Bank Industry and Energy Department, Working Paper, Energy Series 32, Washington D.C.: World Bank

Meyers, S.; Goldman, N.; and Friedman, M. (1993) "Prospects for the power sector in nine developing countries", *Energy Policy* 21 1123–1131

Miller, D. (1998) 'Agents of sustainable technological change: the role of entrepreneurs in the diffusion of solar photovoltaic technology in the developing world', unpublished Ph.D. thesis, Judge Institute of Management, University of Cambridge

Munasinghe, M. (1990) *Electric power economics*, London: Butterworth

Nonaka, I. and Takeuchi, H. (1995) *The knowledge-creating company: how Japanese companies create the dynamics of innovation*, Oxford: Oxford University Press

O'Riordan, T, and Jäger, J. (eds) (1996) *Politics of climate change: a European perspective*, London: Routledge

O'Riordan, T. and Jordan, A. (1996) 'Social institutions and climate change', pp.65–105 in O'Riordan, T, and Jäger, J. (eds) *Politics of climate change: a European perspective*, London: Routledge

OECD (1997) *Globalisation and environment: preliminary perspectives*, Paris: OECD

Ohkawa, K. and Otsuka, K. (1994) *Technology diffusion, productivity employment, and phase shifts in developing countries*, Tokyo: University of Tokyo Press

Parikh, J. (1995) "Joint implementation and North–South cooperation for climate change", *International Environmental Affairs* 7:1 22–41

Pasuk Phongpaichit and Baker, C. (1995) *Thailand: economy and politics*, Oxford: Oxford University Press

Patterson, W. (1999) *Transforming electricity: the coming generation of change*, London: RIIA and Earthscan

Peattie, K. (1995) *Environmental marketing management* London: Pitman

Peralta, D. (1993) 'Energy policy in the Philippines and the proposed privatisation programme', pp.2.1–2.5 in *Asian electricity: the growing commercialisation of power generation*, FT Conference, Singapore 25–26 May 1993, Wolverhampton: The Freelance Association

Penrose, E. (1959) *The theory of the growth of the firm*, Oxford: Oxford University Press

Pitelis, C. (1991 paperback edition, 1993) *Market and non-market hierarchies: theory of institutional failure*, Oxford: Blackwell

Pitelis, C. (ed) (1993) *Transaction costs, markets and hierarchies*, Oxford: Blackwell

Porter, M. (ed) (1986) *Competition in global industries*, Boston, Mass.: Harvard Business School Press

Porter, M. (1990) *The competitive advantage of nations*, New York: Free Press

Porter, M. and van der Linde, C. (1995) 'Green and competitive: ending the stalemate', *Harvard Business Review* September –October: 120–134

Quinn, B. (1977) 'Creating electricity markets in transition economies: the case of Vietnam', paper presented at the IAEE conference, San Francisco, November 1997

Ramani, K.; Hills, P.; and George, G. (eds) (1992) *Burning questions: environmental limits to energy growth in Asian-Pacific countries during the 1990s*, Kuala Lumpur: Asian and Pacific Development Centre

Ramani, K.; Islam, M.; and Reddy, A. (eds) (1993) *Rural energy systems in the Asia-Pacific: a survey of their status, planning and management*, Kuala Lumpur: Asian and Pacific Development Centre

Ramani, K. (1997) 'Rural electrification and rural development', paper submitted to the Asian and Pacific Initiative for Renewable Energy conference, Jakarta 14–16 October 1997

Rakwichian, W.; O'Donoghue, J.; and Na-Lamphun, B. (1996) 'Tapping renewable resources: a new organisation helps to promote sustainable energy in the Mekong region', *Gate (German Appropriate Technology Exchange)*, 4/1996 pp.16–20

Ranganathan, V. (ed) (1992) *Rural electrification in Africa*, London: Zed Books

Rayner, S. and Malone, E. (eds) (1998) *Human choice and climate change: an international assessment*, Washington D.C.: Battelle Press

Reich, R. (1991) *The work of nations: preparing ourselves for 21st–century capitalism*, London: Simon and Schuster

Reid, W. and Goldemberg, J. (1997) 'Are developing countries already doing as much as industrialised countries to slow climate change?', *World Resource Institute Climate Notes*, July 1997

Reimer, P., Smith, A. and Thanbimuthu, K. (eds) (1997) *Greenhouse gas mitigation: technologies for activities implemented jointly,* Pergamon, Elsevier Science:m Oxford

RENI (1997), *Renewable energy in Indonesia: trade guide*, Jakarta: Renewable Energy Network Indonesia, Winrock International, and Yayasan Bina Usaha Lingkungan

Rigg, J. (ed) (1995) *Counting the costs: economic growth and environmental change in Thailand*, Singapore: ISEAS

Riordan, M. (1990) 'What is vertical integration?' pp.94–111 in Aoki, M.; Gustafsson, B. and Williamson, O. (eds) *The firm as a nexus of treaties*, London: Sage

References

Ripple, D. and Takahoshi, E. (1997) *Independent power producers in Asia: past, present and future*, London: FT Energy

Rosenzweig, M. and Voll. S. (1997) 'Sequencing power sector privatisation: is reform its precondition or result?', *Pacific and Asian Journal of Energy* 7:2 119–132

Sathaye, J. and Tyler, S. (1991) "Transitions in household energy use in urban China, India, the Philippines, Thailand and Hong Kong", *Annual Review of Energy and the Environment* 16 295–335

Steiger, J. (1988) *Renewable energy sources in ASEAN*, ASEAN Economic Research Unit, Research Notes and Discussions Paper no.64, Singapore: ISEAS

Stobaugh, R. and Wells, L. (eds) (1984) *Technology crossing borders: the choice, transfer and management of international technology flows*, Boston: Harvard Business School

Stone, A.; Levy, B.; and Paredes, R. (1996) 'Public institutions and private transactions: a comparative analysis of the legal and regulatory environment for business transactions in Brazil and Chile', pp.95–128 in Alston, L.; Eggertsson, T.; and North, D. (eds) *Empirical studies in institutional change*, Cambridge: Cambridge University Press

Stoneman, P. (1987) *The economic analysis of technology policy*, Oxford: Oxford University Press

Symon, A. (1997) *Energy in Indonesia*, London FT Energy

TEI (Thailand Environment Institute) (1997) *Climate technology initiative (CTI related survey)*, submitted to the International Centre for Environmental Technology Transfer (ICETT), Japan, March 1997

TERI (Tata Energy Research Institute) (1997) *Capacity building for technology transfer in the context of climate change*, New Delhi: TERI

Tidd, J.; Bessant, J. and Pavitt, K. (1997) *Managing innovation: integrating technological, market and organisational change*, Chichester, UK: Wiley

Toan, Pham Khanh (1997) 'Renewable energy in Vietnam', paper presented at the *Asia Pacific Initiative for Renewable Energy and Energy Efficiency Conference*, 14–16 October 1997, Jakarta

Topping, J.; Qureshi, A. And Dabi, C. (1996) "Building on the Asian Climate Initiative: a partnership to produce radical innovation in energy systems", *Journal of Environment and Development* 5:1 4–27

UNCTAD (United Nations Comission on Trade and Development) (1997) "Promoting the transfer and use of environmentally sound technologies: a review of policies", *Science and Technology Issues*, Geneva: UNCTAD

US Agency for International Development (USAID) (1990) *Greenhouse has emissions and the developing countries: strategic options and the USAID response*, A report to Congress, Washington D.C.: USAID.

US Agency for International Development (USAID) (1998) *The environmental implications of power sector reform in developing countries*, A report for the Office of Energy, Environment, and Technology, Washington D.C.: USAID.

US Department of Energy (1997) *Scenarios of US Carbon reductions: potential impacts of energy technologies by 2010 and beyond*, US Department of Energy: Washington DC

Ventkaratnam, M. (1984) *Dynamics of electrical energy for rural development, the Indian experience*, New Delhi: Rural Electrification Corporation

von Moltke, K. (1992) "International trade, technology transfer and climate change", pp.295–304 in Mintzer, I. (Ed) *Confronting climate change: risks, implictions and responses*, Cambridge: Cambridge University Press

Wade, R. (1990) *Governing the market: economic theory and the roleof government in East Asian industrialisation*, Princeton: Princeton University Press

Walden, D. (1997) 'Business development plan for biomass to energy projects in Indonesia', paper presented at the *Asia Pacific Initiative for Renewable Energy and Energy Efficiency Conference*, 14–16 October 1997, Jakarta

Wallace, D. (1995) *Environmental policy and industrial innovation: strategies in Europe, the US and Japan*, London: Earthscan/RIIA

West, P. (1984) *Foreign investment and technology transfer: the tire industry in Latin America*, London, Greenwich Ct.,: JAI Press

Wigley, T.; Richels, R. and Edmonds, J. (1996) 'Economic and environmental choices in the stabilization of atmospheric CO_2 concentrations', *Nature* 379:6582, 18 January

World Energy Council (1994) *New renewable energy resources: a guide to the future*, London: Kogan Page

Williamson, O. (1975) *Markets and hierarchies: analysis and antitrust implications*, New York: Free Press

Williamson, O. (1985) *The economic institutions of capitalism: firms, markets, relational contracting*, New York: Free Press

Williamson, O. (1990) 'The firm as a nexus of treaties: an introduction', pp.1–25 in Aoki, M.; Gustafsson, B. and Williamson, O. (eds) *The firm as a nexus of treaties*, London: Sage

Williamson, O. (ed) (1995) *Organisation theory: from Chester Barnard to the present and beyond*, Oxford: Oxford University Press

Williamson, O. (1996) *The mechanisms of governance*, Oxford: Oxford University Press

Wiser, R. (1997) 'Renewable Energy Finance and Project Ownership', *Energy Policy*, 25:1 15–27

Woravech, P. (1997) 'NRSE programs in Thailand', paper presented at the *Asia Pacific Initiative for Renewable Energy and Energy Efficiency Conference*, 14–16 October 1997, Jakarta

World Bank (1991) *Core report of the electric power efficiency improvement study*, Industry and Energy Department, Washington D.C.: World Bank

World Bank (1993) *Energy efficiency and conservation in the developing world: the World Bank's role*, World Bank Policy Paper, Washington D.C.: World Bank

World Bank (1995) *Infrastructure development in East Asia and Pacific*, Washington, D.C.: World Bank

World Bank (1996) *Rural energy and development: improving energy supplies for two billion people*, Washington D.C.: World Bank

World Bank, Asian Development Bank, Food and Agriculture Organisation, and Water Resources Group, in cooperation with the Institute of Water Resources Planning, Vietnam (1996) *Vietnam: water resources sector review: main report*, Hanoi: World Bank

Wu, Changqui (1992) *Strategic aspects of oligopolistic vertical integration*, Studies in mathematical and managerial economics 36, Amsterdam: North Holland

Wynne, B. (1994) 'Scientific knowledge and the global environment', in Redclift, M. and Benton, T. (eds) *Social theory and the global environment*, London: Routledge

Yamashita, S. (ed) (1991) *Transfer of Japanese technology and management to the ASEAN countries*, Tokyo: University of Tokyo Press

Yearley, S. (1996) *Sociology, environmentalism, globalization*, London: Sage

Yuen, Ng Chee; Freeman, N.; and Huynh, F. (eds) (1996) *State owned enterprise reform in Vietnam: lessons from Asia*, Singapore: Institute of Southeast Asian Studies